Pesticides - Balancing Efficient Pest Control with Environmental and Health Safety

Edited by Ravendra Kumar,
Sonu Kumar Mahawer
and Mozaniel Santana de Oliveira

Published in London, United Kingdom

Pesticides - Balancing Efficient Pest Control with Environmental and Health Safety
http://dx.doi.org/10.5772/intechopen.1008610
Edited by Ravendra Kumar, Sonu Kumar Mahawer and Mozaniel Santana de Oliveira

Contributors
Adeola A. Oso, Aydemir Bariş, Derya Kocamaz, Feng Liu, Jia Xing Huang, Kedi Zhao, Lifeng Meng, Li Jiao Gao, Oguz Merhan, Qiancheng Wei, Samuel C. Spengler, Steven R. Spengler, Tarek Othman Said, Xiaojun Wang, Yiying Zhao

First published in London, United Kingdom, 2026 by IntechOpen
IntechOpen is the global imprint of INTECHOPEN LIMITED, registered in England and Wales, registration number: 11086078, 167-169 Great Portland Street, London, W1W 5PF, United Kingdom

For EU product safety concerns: IN TECH d.o.o., Prolaz Marije Krucifikse Kozulić 3, 51000 Rijeka, Croatia, info@intechopen.com or visit our website at intechopen.com.

British Library Cataloguing-in-Publication Data
A catalogue record for this book is available from the British Library

Pesticides - Balancing Efficient Pest Control with Environmental and Health Safety
Edited by Ravendra Kumar, Sonu Kumar Mahawer and Mozaniel Santana de Oliveira
p. cm.
Print ISBN 978-1-83769-611-6
Online ISBN 978-1-83769-610-9
eBook (PDF) ISBN 978-1-83769-657-4

If disposing of this product, please recycle the paper responsibly.

Meet the editors

Prof. Dr. Ravendra Kumar is an Assistant Professor in the Department of Chemistry, G. B. Pant University of Agriculture & Technology, Pantnagar, India. He has published over 135 research articles, 22 book chapters, and edited 3 books, including Biorationals and Biopesticides (de Gruyter, 2024) and Natural Pesticides and Allelochemicals (CRC Press, 2024). His research focuses on bioactive natural plant products, including essential oils, terpenes, flavonoids, and alkaloids, as well as nanoformulations targeting insects, nematodes, and fish pathogens. He holds an Indian patent for a portable RP-HPLC device and leads national projects funded by UCB and UCOST. Dr. Kumar has received multiple awards, including the Radhakrishnan Best Teacher-Researcher Award (2025) and Excellence in Research Awards (2023, 2025).

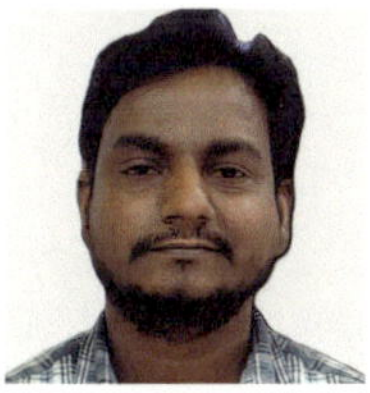

Dr. Sonu Kumar Mahawer is a Scientist in Agricultural Chemicals at the Division of Crop Production, ICAR-Indian Grassland and Fodder Research Institute, Jhansi, India. He earned his Ph.D. in Agricultural Chemicals from Govind Ballabh Pant University of Agriculture & Technology, Pantnagar, in 2024. Dr. Mahawer has published over 42 research and review articles, 22 book chapters, and more than 15 popular articles in national and international journals. His research focuses on pesticide residue chemistry, natural product chemistry, fodder chemistry, and contaminants. He has qualified for ICAR-NET and ARS and continues to advance knowledge in agricultural and natural product chemistry through both research and applied projects.

Dr. Mozaniel Santana de Oliveira holds a Ph.D. in Food Science and Technology from the Federal University of Pará (UFPA, 2018), where he also completed his Bachelor's (2014) and Master's (2016). He conducted postdoctoral research in Tropical Botany at UFRA/MPEG (2023–2025). A specialist in natural products chemistry, he has expertise in essential oils, secondary metabolite extraction, and modern analytical techniques. Dr. Oliveira has authored 266 publications, including 130 international journal articles, 14 books, and 60 book chapters, and has reviewed over 667 manuscripts for around 90 journals. His research focuses on bioactive molecules, pharmacological applications, allelopathy, and food biotechnology. He currently serves as a Visiting Professor in Pharmaceutical Sciences at UFPA and maintains active international collaborations.

Contents

Preface

The growing demand for agricultural productivity and global food security has reinforced the strategic importance of pesticides in modern farming systems. At the same time, heightened awareness of their environmental persistence, ecological disruption, and potential human health risks has intensified the search for safer, more sustainable alternatives. The central purpose of this edited volume is to present a balanced, scientifically grounded discussion of how efficient pest control can coexist with environmental protection and health and safety. This book brings together multidisciplinary perspectives that address both the challenges and innovations shaping contemporary pesticide science.

It opens with "Challenges and Future Orientation of Pesticides", which outlines the limitations of conventional pesticide use, discusses regulatory and environmental pressures, and highlights the need for technological transformation toward safer and more targeted solutions. This chapter establishes the conceptual framework for the volume.

The second chapter, "Pesticides: Balancing Efficient Pest Control with Environmental and Health Safety," explores the critical balance between effective pest control and the protection of human health and the environment, highlighting key risks, regulatory considerations, and sustainable management strategies.

The third chapter, "Pesticide Toxicity to Pollinators: Impacts on Bee Growth, Development, and Behavior", explores the ecological risks posed by pesticide exposure, particularly for pollinators. By examining developmental, physiological, and behavioral effects, it underscores the importance of incorporating sublethal and long-term impacts into environmental risk assessments.

Environmental contamination is further examined in "Anthropogenic Impact to Stream and Groundwater Quality on the Island of Oahu, Hawaii", which analyzes how agricultural and human activities influence water quality. This chapter provides valuable insights into the transport, persistence, and ecological implications of pesticide residues in aquatic systems.

In response to environmental concerns, "Environmental Perspectives on Biopesticides: Current Knowledge and Research Gaps" evaluates biologically derived pest control agents as promising alternatives to conventional chemicals. The discussion highlights current advances while identifying critical knowledge gaps that must be addressed to strengthen the reliability and regulatory acceptance of biopesticides.

Emerging molecular strategies are explored in "RNA Interference Technology: From Molecular Mechanisms to Applications as Green Pesticides", which presents RNA interference (RNAi) as a highly specific and innovative tool for pest control. The chapter discusses its mechanistic basis and its potential to minimize off-target effects and environmental persistence.

The volume concludes with "The Melon Fly (Myiopardalis Pardalina) (Diptera: Tephritidae): Identification, Biology, Damage Characteristics, and Management Strategies", offering a detailed case study of a significant agricultural pest. By integrating biological knowledge with management approaches, it illustrates how targeted strategies can reduce reliance on broad-spectrum chemical pesticides.

These chapters highlight that the future of pesticide science depends on integration: combining ecological awareness, technological innovation, regulatory responsibility, and sustainable agricultural practices. This volume takes a balanced approach, showing how pesticides fit into the larger picture of environmental and public health.

As Academic Editor, I extend my sincere appreciation to all contributors and reviewers whose expertise and dedication made this volume possible. We hope this book will serve as a valuable reference for researchers, environmental scientists, agronomists, toxicologists, policymakers, and students seeking informed, balanced approaches to pest management.

Ravendra Kumar
Department of Chemistry at the College of Basic Sciences and Humanities,
G. B. Pant University of Agriculture and Technology,
Pantnagar, Udham Singh Nagar, Uttarakhand, India

Sonu Kumar Mahawer
Crop Production Division,
ICAR-Indian Grassland and Fodder Research Institute,
Jhansi, Uttar Pradesh, India

Mozaniel Santana de Oliveira
Institute of Health Sciences at the Federal University of Pará (UFPA),
Belém, Brazil

Chapter 1

Challenges and Future Orientation of Pesticides

Tarek Othman Said

Abstract

Safety measures are desperately needed to lower the health dangers linked to contemporary pesticides. How successfully the current laws protect the environment and public health is one of the primary issues. The scientific community can help develop effective, empirically based policies and practices that protect the environment and public health, increase our understanding of the consequences of pesticides, and encourage their ethical use. Furthermore, in order to safeguard human health and preserve a pollution-free environment, international pesticide monitoring criteria frequently need to be modified in light of new scientific discoveries. To develop guidelines and rules for the creation of pesticides with fewer negative effects, many sectors, including manufacturers, government agencies, and non-governmental groups, should work together. Understanding these impacts can improve decision-making, reduce risks, enhance safety procedures, practice environmental stewardship, safeguard community health, and facilitate cooperation and communication among stakeholders. Numerous factors, such as consumer preferences, legislative changes, technological advancements, and a greater understanding of the significance of sustainable agriculture, are anticipated to influence pesticides in the future. In conclusion, it is critical to strike a balance between the necessity of pest management and the potential risks associated with pesticides.

Keywords: pesticides, health risks, challenges, future prospects, orientation

1. Introduction

There are many possible problems associated with pesticides, such as health risks, environmental damage, and pests that are resistant to pesticides [1, 2]. To protect people and the environment, a variety of actions can be performed. For example, governments, research institutions, and private companies can invest in the creation of safer alternatives to chemical pesticides. Integrated pest management (IPM) emphasizes using multiple pest management techniques to reduce the use of pesticides [3]. This entails spotting and identifying pests, setting action thresholds, and controlling them mechanically or physically. Tight rules governing the use and sale of pesticides should be implemented and monitored by regulatory bodies. It is necessary to conduct thorough testing and risk assessments before approving the use of any pesticide for commercial purposes. Regular monitoring and assessment should also be carried out to ensure that safety regulations are being observed [4].

Furthermore, instruction and training on the correct handling, storage, and application of pesticides should be provided, particularly to farmers, agricultural workers, and pesticide applicators. This means using protective gear, following label instructions, and being aware of potential risks and safety measures. International cooperation and coordination between countries are crucial to addressing the hazards that pesticides pose on a global scale [5].

1.1 Harmful consequences on the ecosystem

Pesticides are often used to protect crops, increase yields, and combat parasite illnesses during disasters. In addition to being a major hazard to life and the environment, pesticides can spread through leaching and runoff [6–8]. Specialists, the general population, and institutions frequently use and dispose of pesticides, which can expose the environment. It is difficult to completely control their distribution and the risks they pose to humanity since they can come from a variety of sources, including soil, water, and air [9]. Pesticides, particularly those with long half-lives, can remain in the soil for many years. When pesticides are used excessively, hazardous materials can accumulate in the soil and harm beneficial species like earthworms as well as microbes. This can negatively impact soil fertility and ecosystem health [10, 11]. Pesticides used in agriculture have the ability to percolate into groundwater or reach adjacent bodies of water through surface runoff. Pesticides have diverse environmental impacts, ranging from minor disruptions to species extinction. Pesticides can cause both immediate and long-term harm.

1.1.1 Case studies

It was discovered that over 70 distinct pesticide residues were present in top soils throughout Europe; glyphosate and its metabolites were regularly found to be present in the greatest amounts. Pesticides found in Europe's rivers, lakes, and surface water may harm aquatic species [12, 13]. The largest danger to aquatic life is pesticide contamination, which lowers dissolved oxygen levels. They impact fish and algae as well as aquatic animals at all trophic levels. Fish populations may be declining as a result of pesticides discovered in surfaces and water [14, 15]. Aquatic animals consume pesticides in various ways, such as through skin contact, inhalation, and ingestion. Herbicides reduce oxygen levels in aquatic environments, leading to fish suffocation and decreased output. Herbicide use near weedy fish nurseries has negatively impacted aquatic species' reproductive capacities [14]. By altering soil chemistry, interacting with plant roots, and reducing nitrogen fixation by rhizobiological bacteria, pesticides have an impact on soil microflora, microfauna, and macrofauna [16]. An Indian survey found that 58% of drinking water samples from hand pumps and wells close to Bhopal had organochlorine levels over the US EPA threshold. Pesticides were found in over 90% of fish and aquatic samples from Indian streams. Indian drinking water was found to contain significant amounts of pesticides and their derivatives, including DDT metabolites like hexachlorocyclohexane and endosulfan [17–19]. Pesticides significantly affect plant and animal biodiversity in addition to non-target creatures. Pesticides can affect wildlife indirectly by building up in the food chain. Raptors and mammals, which are at the top of the food chain, may be exposed to high levels of pesticides via contaminated prey. This could lead to reduced immune systems, decreased fertility, and population declines [20, 21]. In general, those with compromised immune systems are more susceptible to illnesses such as influenza, pneumonia, and bronchitis [22, 23].

Pesticides may contaminate natural ecosystems in two ways, depending on their solubility. Pesticides can end up in rivers, lakes, streams, and groundwater after dissolving in water, where they might injure animals that are not their intended target. Bioamplification disrupts the entire ecosystem, causing higher trophic-level animals to die from increased toxicity in their bodies. Consequently, the number of secondary consumers has increased while the number of primary consumers has declined [24]. Targeted pests may become resistant to pesticides following extensive and prolonged treatment. It might take higher dosages or more toxic choices to get the desired result when pests develop resistance to a pesticide. The adverse effects on the ecosystem may develop as a result of increased pesticide use [25, 26]. The effects of pesticides on people and the environment are shown in **Figure 1**.

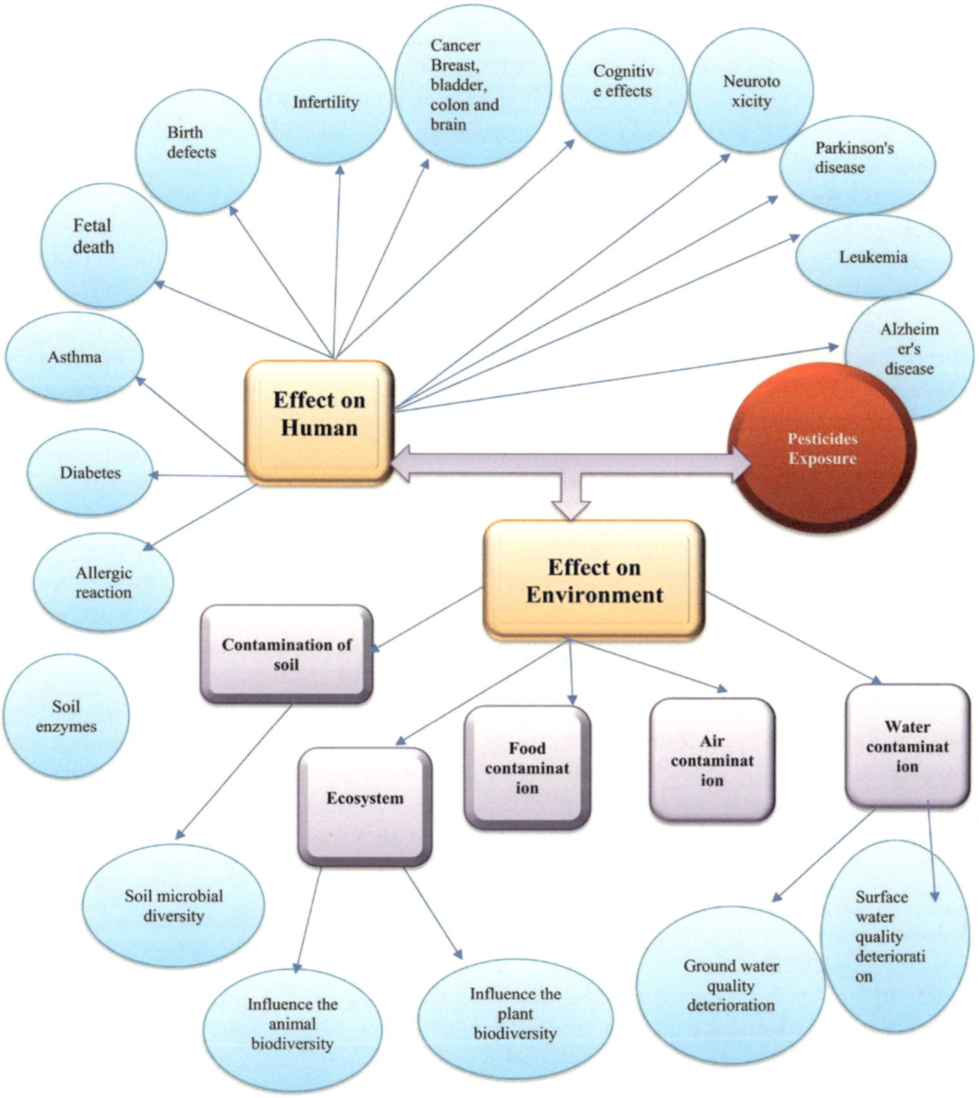

Figure 1.
The effects of pesticides on humans and the environment [32].

2. Strategies for safety precautions

In the contemporary environment affected by pesticides, safety and health risks are major problems. Numerous regulatory agencies oversee the safety of pesticides. Experts in agriculture have begun creating other strategies to mitigate the harm caused by agricultural chemical practices. Integrated crop management (ICM) provides guidelines for farmers to adhere to when implementing policies to produce safe agricultural goods while also being ecologically sensitive [27–30]. ICM also covers protocols for implementing specific environmental preservation measures, worker safety and hygiene, product security, full measurement traceability, and good agricultural practices (GAPs) [31].

ICM promotes the adoption of complementary pest management techniques, such as crop tolerance to fungus and insects, biological control, and safety, while reducing the negative effects of pesticides on other agro-ecosystem components [33, 34]. ICM allows the use of pesticides only through an IPM program, where specific criteria are used to select pesticides, detailed instructions are followed when applying them to crops, and residue analysis is used as one of the enforcement tools [30, 31]. The selection of pesticides for IPM is influenced by the following factors: (1) biological efficiency (plant tolerance, rapid impact, optimal residual effect, reduced resistance risk, and great selectivity); (2) ease of use (lower toxicity, optimal formulation, safe packaging, long shelf life); (3) environmental friendliness (low soil movement, quick degradation in the atmosphere); and (4) financial advantages [35]. It is evident that installing an IPM system would help decrease the negative impacts of pesticides without affecting agricultural output or increasing the risk of crop losses [27]. Furthermore, it is vital to encourage cooperation and communication among all stakeholders involved, including farmers, workers, researchers, regulatory bodies, and pesticide applicators. Such initiatives can support the creation of novel and safer methods. Emphasis should be placed on choosing less hazardous pesticides with lower toxicity and environmental impact. The use of non-chemical substitutes should be promoted whenever feasible, taking into account physical barriers, biological limitations, or cultural norms [36, 37]. Many nations have implemented occupational safety laws and regulations to shield workers from pesticide exposure. These laws frequently require businesses to provide suitable personal protective equipment (PPE), instruction on the safe handling and application of pesticides, and ongoing health monitoring for staff members [38, 39]. In order to efficiently manage pests and reduce dependency on chemical pesticides, IPM also emphasizes preventive methods, including enhancing sanitation and using biological controls. Employee exposure levels to pesticides must be regularly monitored and documented to identify any problems and take the necessary action.

3. Laws pertaining to pesticides

Handford et al. [40] examined pesticide laws around the world and found notable regional and national variations. They observed that impoverished nations lack the means and expertise to enforce laws, while industrialized ones often have stricter regulations. Specifically, the European Union (EU) is renowned for having some of the strictest pesticide regulations in the world [40]. During a significant review of all active compounds allowed for use in the EU between 1998 and 2009, three-quarters of all active chemicals that were previously licensed for use in the EU were banned. Maximum residue limits (MRLs) are used in the majority of countries to regulate pesticides. MRLs are the highest level of a pesticide residue that is legally tolerated in or on food or feed when pesticides are applied correctly. MRLs typically vary significantly between nations [40]. When comparing the MRLs of the United States, Canada, China, India, Japan, South Africa, and Australia, the United States had the highest MRLs, while the EU had the lowest (signaling the strongest regulations). In contrast to the EU, developing nations lack sufficient legislation to properly regulate the use, storage, labeling, disposal, and transportation of pesticides due to a lack of resources and expertise to increase regulation, as well as a lack of awareness of the risks involved. In Brazil, for example, three agencies – agriculture, which records and analyzes goods; health, which reviews human health hazards; and environment, which evaluates environmental threats – share responsibility for pesticide laws. Pesticide-regulating agencies differ from country to country. Japan is one of the largest pesticide users in the world. The Japanese Ministry of Health has authorized MRLs [40, 41]. The Institute for the Control of Agrochemicals was formed by the Ministry of Agriculture to enforce Chinese law. Manufacturers must submit a comprehensive application for novel compounds and uses for pesticides that are already registered to the US Environmental Protection Agency (USEPA), which is responsible for pesticide registration. Nonetheless, scholars and reporters have discussed whether the laws in place are sufficient to safeguard both the environment and human health [42, 43]. In the meantime, Brazil is the biggest consumer of pesticides worldwide [44, 45].

4. Future potential

Pesticides pose concerns such as health hazards, environmental damage, and pest resistance [1, 2]. To protect human and environmental safety, governments, research groups, and private firms can invest in developing safer alternatives to chemical pesticides. To use fewer pesticides, IPM combines multiple pest management strategies [45], including pest monitoring, identification, action thresholds, and other control procedures. Pesticide usage and sales should be limited by rigorous regulations enforced through integrated monitoring. Extensive testing and risk assessments are required before commercial pesticide use is approved. Oversight and assessment should be conducted to ensure that safety criteria are met [4]. Furthermore, those who apply pesticides should be educated and trained on correct pesticide handling, storage, and application techniques. This includes understanding potential hazards and safety precautions, using protective equipment, and following label requirements. To counter the global threat posed by pesticides, governments must coordinate and cooperate [5].

5. Scenarios for sustainable pesticide use

For over 45 years, the EU has prioritized sustainable pesticide usage through several laws and regulations. Agrochemical companies, lawmakers, farmers, and residents are among the stakeholders who pose challenges to achieving EU pesticide reduction goals and minimizing their hazards. A foresight investigation was commissioned by the European Commission's Directorate General for Health and Food Safety (DG SANTE) to examine pesticide use in 2030 and beyond, with an emphasis on reaching specific pesticide-use and risk-reduction goals. Scenarios are a method for analyzing uncertainty and representing potential future outcomes. Scenarios are not forecasts but rather offer a variety of potential future situations for exploring policy consequences. This method examines possible outcomes for 2030 and beyond. Given the intricate structure of the policy domain and the numerous factors influencing its creation and implementation, this strategy acknowledges the possibility of minimal change by 2030 [46]. Numerous factors, such as political, economic, and sociological ones, as well as technological and environmental ramifications, influence the use of pesticides in the EU. To predict the future by 2030 and beyond, it is important to evaluate the trends, interdependence, and uncertainty that shape these aspects.

5.1 Scenario 1: Various sustainable approaches

In response to geopolitical instability and climate change, the EU is taking an inward-looking approach to strengthen its agricultural sector. The agrifood sector is gradually adopting more sustainable practices. The usage of chemical pesticides has decreased somewhat, but the trend has not been uniform across the agriculture sector. Financial and policy support are required to encourage innovation and alternative methods to further minimize the application of pesticides [46].

5.2 Scenario 2: Business sustainability

The EU's economy has grown moderately, while agricultural prices have been largely steady. Change is occurring more slowly than anticipated by market estimates, notwithstanding consumers' progressive shift toward diets with less of an impact on the environment. In the absence of substantial legislative and regulatory incentives from the EU, large firms are increasingly driving sustainability and innovation in the agriculture sector. Generally speaking, the EU has been using fewer chemical pesticides. It still has an impact, though, because farmers are not always willing to adopt new methods (like IPM) to cut back on or completely stop using pesticides.

5.3 Scenario 3: Unsustainable approach

Innovation and sustainable practices have lost political traction as a result of the global recession. This has resulted in a lack of alternatives for pest management, a lack of funding for farmers to implement innovative strategies, and a persistent dependence on chemical pesticides. Affordability is more important to consumers than environmental sustainability. In the global agrifood system, the dominance of huge businesses across the food supply chain is displacing small-scale producers. Climate change has not yet had a significant effect on agricultural production, which has remained stable in terms of yield, pricing, and competitiveness [46].

5.4 Scenario 4: Pervasive sustainability

Pesticides pose major risks, even though they are necessary for agricultural productivity and health. To address these issues, future possibilities call for a move toward safer substitutes, accurate application, and environmentally friendly procedures. The EU policies show a greater commitment to addressing environmental challenges and promoting sustainable economic growth. These goals are thought to be significantly facilitated by technologies and alternative farming methods that allow for the elimination of pesticides, which has led to their extensive implementation throughout the food supply chain. Additionally, EU consumers are becoming more environmentally conscious. These developments have facilitated a swift transition to ecologically friendly methods, such as cutting back on the usage of pesticides. The willingness of consumers to pay extra for goods made using better production methods, higher environmental and health regulations, and more innovative technology that promotes sustainable practices has lessened the impact of rising food prices [46].

6. Risk-reduction targets

The EU establishes a number of risk-reduction goals, including a 50% decrease in the usage and risk of pesticides and more dangerous chemicals by 2030. These non-binding goals are a component of the EU's larger initiative to shift to a more ecologically friendly and sustainable food system [47].

7. Biodiversity strategies

The 2030 biodiversity strategy of the EU is a comprehensive, long-term, and ambitious plan to protect the environment and halt ecological degradation. By 2030, the strategy aims to put EU biodiversity on a path to recovery through specific actions and commitments. It outlines the EU's strategy for participating in the upcoming round of international discussions on the framework for biodiversity beyond 2020. It is an essential part of the EU accord and helps promote a green rebound following the COVID-19 pandemic [47].

7.1 Goals

For the benefit of its people, the environment, and the climate, the biodiversity strategy aims to restore Europe's biodiversity by 2030. The strategy's goal in the post-COVID-19 environment is to increase our societies' resilience to potential dangers like the effects of food instability, forest fires, and disease outbreaks due to climate change, especially by preventing the illegal wildlife trade and preserving animals [47].

7.2 Actions

Specific commitments and actions that must be fulfilled by 2030 are outlined in the EU strategy. The EU is creating a wider network of marine and land-based protected areas. They will increase the number of existing Natura 2000 areas, ensuring that areas of exceptional biodiversity and climate significance are rigorously safeguarded [48].

7.2.1 Establishing a plan for EU nature restoration

The plan urges EU nations to implement efficient restoration measures through tangible promises and actions in order to prevent and lessen the effects of natural disasters and restore damaged ecosystems, particularly those with the greatest potential to capture and store carbon. The first-ever law in the EU includes enforceable restoration targets for particular species and ecosystems, a broad restoration goal for the long-term recovery of nature in EU land and marine areas, and legislation to enable the required drastic transformation. The strategy places a strong emphasis on allocating funds for biodiversity and creating a new, improved governance structure to monitor development, guarantee better execution, and expand environmental awareness, resources, and investments further when making corporate and public decisions [48].

7.3 Introducing actions to address the problem of global biodiversity

The EU will demonstrate that it is ready to lead by example in tackling the danger to biodiversity worldwide by implementing these measures, in particular, aiming for the Convention on Biodiversity to effectively execute a comprehensive global scope for biodiversity [48].

8. Implementations

Two web-based resources monitor the strategy's implementation status. An online activities tracker offers the most recent data on how well the strategy's numerous actions are being carried out. An objectives dashboard displays the EU's and the Member States' progress toward the strategy's quantified biodiversity targets [47, 48]. Long-term exposure to pesticide residues can cause persistent soil pollution, which can damage microbial populations and endanger high trophic levels, including plants and animals that rely on healthy soil ecosystems [49]. Non-target species may suffer as a result of pesticide bioaccumulation in the food chain. This process disrupts the environment's balance and endangers biodiversity by negatively affecting organisms that consume polluted plants or soil species [50]. A crucial area of attention is the convergence of public health and policy initiatives to lower the use of pesticides in agriculture, especially with regard to IPM tactics. IPM is well known for being a sustainable strategy that uses a variety of pest management methods, such as biological controls, cultural practices, and the targeted use of pesticides. According to Lopez-Antia [51], this integrated method ensures agricultural yield while minimizing environmental impacts. Reducing the use of pesticides is crucial, and wherever possible, alternatives are encouraged by European Directive 128/2009/EC, which guarantees that decisions about pesticide application are made based on economic considerations [51]. The larger objectives of sustainable agriculture, which seek to strike a balance between ecological integrity and agricultural productivity, are supported by this directive. An outline of the consequences of lowering pesticide use in agriculture, public health initiatives, and alternative farming methods is given in **Table 1**.

Category	Implications
Lowering the usage of pesticides in farming	Increased soil fertility and biodiversity through organic and sustainable practices, better ecosystem health through biological and precision agriculture, and enhanced sustainability through the application of IPM
Campaigns for public health	Increased farmer safety through instruction on pesticide application techniques; decreased consumer exposure through safe food handling procedures (washing, cooking, and peeling); and raised consumer understanding of the risks connected with pesticide usage to help them make educated decisions
Encouragement of non-traditional farming methods	Increased use of organic farming lessens reliance; agroecology and permaculture promote healthier ecosystems; and financial incentives encourage farmers to switch to sustainable methods

Table 1.
An outline of the consequences for public health initiatives, encouraging alternative farming methods, and reducing the use of pesticides in agriculture [51].

8.1 Participation of the community and stakeholders

In order to ensure safer food production, especially when it comes to pesticide use and contamination, cooperation between consumers, farmers, and industries is essential (see **Table 2**). Through their purchasing decisions, consumers significantly influence market dynamics, which can incentivize manufacturers to embrace safer and more sustainable methods. For example, many farmers have reduced their use of pesticides due to the increased demand for organic produce, which has reduced the incidence of pesticide residues in food products [52]. This emphasizes how crucial consumer education and awareness are to improving food safety, since well-informed decisions can encourage the adoption of farming methods that put public health and environmental sustainability first [53].

9. Research defects and future

Analyzing research gaps and future objectives in pesticide exposure and its health consequences is essential for agricultural sustainability and public health (see

Category	Actions	Implications
Customers, farmers, and industries roles	Consumers' and producers' cooperation is essential for the production of safe food, and companies are vital in reducing pesticide contamination by promoting substitutes	Reduced usage of pesticides, increased sustainability in agriculture, and improved food safety and quality
Instructional initiatives	Consumer education on food preparation to reduce pesticide residue intake; training programs for farmers on safe pesticide usage and alternatives	Decreased consumer exposure to pesticides; greater understanding and safer farming methods among farmers.
Support for more stringent regulations	The necessity of transparency in food labeling and pesticide residue testing; the role of civil society in promoting stronger pesticide restrictions	Improved environmental health, a stronger regulatory framework, more responsible food safety, and educated customer choices

Table 2.
Potential steps and outcomes of involving stakeholders [53].

Regulations	Direction for the future
Long-term epidemiological research is necessary	Investigate the deep-rooted impacts of low-dosage pesticide exposure, paying particular attention to disadvantaged groups and geographical inequities
New developments in safer insecticides	Create biopesticides and less hazardous substitutes; implement technologies to reduce the use and drift of pesticides
International oversight and control	Boost global collaboration on pesticide residue regulations and standardize maximum residue levels internationally

Table 3.
Research and future limitations in pesticide safety and regulation [54].

Table 3). One of the most pressing needs is the establishment of long-term epidemiological studies that can elucidate the chronic effects of continued low-dosage pesticide exposure. Acute pesticide toxicity is well-established, according to recent research, but chronic effects – especially at low exposure levels – remain little understood. Research has indicated that long-term exposure to pesticides might cause cognitive effects, such as depression and anxiety, especially in agricultural workers [54]. This information gap highlights the need for comprehensive longitudinal studies that track health outcomes over extended periods of time, offering more accurate insights into the long-term effects of pesticide exposure.

There is an urgent need for research on vulnerable populations and regional variations in pesticide exposure and health impacts. Some groups, such as children, pregnant women, and low-income agricultural workers, may be more vulnerable due to socioeconomic conditions and limited access to preventative measures [55]. For instance, in rural parts of underdeveloped countries, inadequate regulatory frameworks and limited access to healthcare often exacerbate pesticide-related health issues [56]. These susceptible groups should be given priority in future studies, which should employ focused epidemiological techniques to evaluate specific hazards and guide customized solutions.

9.1 A plan for the phase-out of pesticides

Numerous professionals and citizens are urging action in response to the growing concern over pesticide use in the EU and its impact. Over a million EU residents have called for a phase-out of pesticides through the Save Bees and Farmers European Citizen Initiative, and they are presently awaiting a response from the European Commission. Furthermore, the Conference on Europe's Future, the Stop Glyphosate ECI, and many Eurobarometer surveys demonstrate how strongly Europeans support significantly reducing pesticide use [57]. Moreover, the scientific community as a whole agrees that failure to limit pesticide use is not an option. More than 6,000 scientists warned in July 2023 that a significant reduction in pesticide use is urgently needed to safeguard both humans and the environment and to guarantee a long-term, sustainable food supply. The EU has yet to address the problem of pesticide reduction, despite the well-established threats that pesticides pose to human health and biodiversity. Despite implementing the sustainable use of pesticides directive (SUD) in 2009, pesticide sales and usage in Europe have not decreased [58]. Since then, as part of the EU's commitment to the Farm to Fork Strategy, they have pledged to reduce pesticide use and the risks they pose by 50% by

2030. However, the sustainable use rule (SUR), which was necessary to accomplish these goals, was repealed in February 2024 [59]. A key component of the dialog's conclusion is the need to transition to more diverse and sustainable agrifood systems that safeguard the environment and the health of both citizens and food producers.

9.2 Biopesticides

Herbicides, algicides, miticides, bactericides, fumigants, termiticides, repellents, insecticides, molluscicides, nematicides, and pheromones are examples of synthetic pesticides that have been employed over time to increase crop yields. In addition, such pesticides harm soil microbes, vegetation, animals, and soil texture. Fish and other aquatic species are frequently killed by excessive pesticide application and their discharge into water bodies during rainfall. Even when fish are still alive, consuming them can cause toxins in the body to biomagnify, leading to deadly diseases like cancer, diabetes, renal disease, liver failure, eczema, neurological damage, cardiovascular disease, and more [60]. The hazards associated with synthetic pesticides have made the use of organic pesticides, also known as biopesticides, which are more affordable, sustainable, and environmentally friendly, vital. Biopesticides can be produced from microbes (metabolites, for instance), plants (exudates, essential oils, and bark, root, and leaf extracts), and biologically generated nanoparticles (such as silver and gold nanoparticles). Microbial pesticides, as opposed to synthetic ones, are easy to obtain, specific in their action, environmentally friendly, and have no long-term impacts. Apart from their diverse array of phytochemical constituents that allow them to exhibit many modes of action, phytopesticides also present lower health risks to people and are not associated with greenhouse gas emissions. Nanobiopesticides have better biocompatibility and biodegradability, targeted or controlled release, and higher pesticidal effect. More importantly, Ayilara et al. [60] examined appropriate and sustainable ways to improve the commercial use and acceptability of phytopesticides, nanobiopesticides, and microbial pesticides for crop protection and yield, plant nutrition, health promotion for humans and animals, and possible integration into the IPM system. The different kinds of pesticides and the benefits and drawbacks of synthetic pesticides and biopesticides were all examined in their review.

9.3 Trends for 2025: Advancements and difficulties in food and agriculture systems

The agriculture industry experienced escalating difficulties in 2024 as a result of geopolitical conflicts, economic uncertainty, and climate change. Farmers' resiliency was put to the test by fluctuating weather patterns and rising input costs. However, as the sector looks to 2025, these difficulties have spurred innovation, leading to technical advancements and sustainable practices. The agricultural sector's response to the 2024 issues demonstrates how important it is to sustain humankind and tackle the mounting demands on the world's nourishment systems [61]. Agriculture and food are the foundations of humanity's existence and economic growth because they supply essential materials for daily life. As the world's population grows, so does the need for food; these systems are under great strain. This issue is made even more urgent by rising inflation and declining food security, underscoring the need for longer-term solutions. Technological advancements are driving growth by offering innovative ways to boost productivity and efficiency. Policies and laws, which are crucial in determining the direction of agriculture, also have a significant impact on these developments.

9.3.1 Rules and policies

Policy and regulation are major drivers of change in the food and agriculture industries. Change is implemented by areas worldwide through updated standards and incentives for innovation and sustainable practices. For example, the USEPA established a number of programs and regulatory actions to lower greenhouse gas emissions, including guidelines that promote the use of carbon-neutral practices and renewable energy sources. Through its energy, transportation, tax, and climate policies, the EU plans to reduce net greenhouse gas emissions by at least 55% by 2030 [61]. Furthermore, the EPA advocates for stronger standards to guarantee food safety and quality. With customers placing a higher priority on food items' quality and safety as a result of these rules, the trend toward food transparency and traceability continues to increase. Advances in blockchain technology, a peer-to-peer network that stores data, and real-time food monitoring through Internet of Things (IoT) devices – a network of physical devices and other objects embedded with sensors, software, and network connectivity that allow them to collect and share data – have made it possible for food brands to offer full end-to-end traceability [62]. These advancements also contribute to a healthier planet and a more transparent supply chain.

9.3.2 Adaptation and sustainable practices

The resilience and sustainability of the food and agriculture industries depend on their ability to adapt to challenges. As extreme weather events increase in frequency and severity, new approaches are needed to mitigate the effects of climate change. Compared to 35% in 2022, 41% of farmers in 2024 identified weather and climate disasters as one of the largest threats to their revenue. Many academics have argued that using climate-resilient crop varieties, such as disease-resistant rice strains, early-maturing soybeans, and drought-tolerant chickpeas, is the most feasible method to lessen the effects of climate change, particularly in vulnerable areas around the world. Using climate-smart agriculture is one approach; this includes, among other things, mulching, using compost or organic fertilizer, rotating crops with legumes, growing nitrogen-fixing trees on the farm to act as windbreaks, and zero or low tillage [63]. Numerous opportunities and challenges are causing significant changes in the food and agriculture industries. Adopting sustainable practices is essential to creating resilient and effective food systems, as extreme weather events and growing input prices are making these changes more challenging. Data-driven solutions and cutting-edge technologies are accelerating this endeavor. Customers' desire for locally produced and sustainable goods is increasingly influencing trends. Maintaining a competitive edge in the market and ensuring a safe and sustainable future for the world's food production depend on keeping up with these changes. The food and agriculture sectors are adopting sustainable practices at a much faster rate because of technological advancements. By enabling more effective use of herbicides, fertilizers, and water, these technologies are helping farmers minimize their negative effects on the environment without sacrificing output. One of the best examples is precision agriculture, which allows farmers to apply precise amounts of inputs to improve crop quality and yield. Satellites and drones are utilized to compute parameters relating to vigor, water stress, and the amount of chlorophyll in order to produce maps that allow the variable rate application of fertilizers and other inputs.

10. Conclusion

Significant threats to the environment and public health are posed by pesticide residues, which can contaminate soil, degrade water, destroy biodiversity, and expose people. A concerted effort from all parties involved is required to address these issues, with an emphasis on creating sustainable farming methods, including organic farming, biological controls, and IPM. Pesticide management in 2025 will be more focused on sustainable methods, with priorities such as encouraging IPM, lowering pesticide use, and striking a balance between environmental preservation and food production. Stricter regulations, research into safer substitutes like biopesticides, and improved farmer education and training on sustainable practices are some of the future directions. To reduce these hazards, it is essential to enhance research into safer pesticides and strengthen regulatory systems. Public health initiatives, such as awareness-raising campaigns and more transparent food labels, can empower consumers and lower exposure, particularly for vulnerable groups. Promoting community-driven change and collaboration among farmers, companies, and consumers can help facilitate the transition to safer food production systems. We can create a more resilient and healthier agricultural system that safeguards human health and the environment if we give these projects top priority.

Declaration of Competing Interest

The author declares that there are no competing financial interests or personal relationships that could have appeared to influence this work.

Author details

Tarek Othman Said
Professor of Marine Chemistry, National Institute of Oceanography and Fisheries (NIOF), Cairo, Egypt

*Address all correspondence to: tareksaideg@yahoo.co.uk

References

[1] Cech R, Zaller JG, Lyssimachou A, Clausing P, Hertoge K, Linhart C. Pesticide drift mitigation measures appear to reduce contamination of nonagricultural areas, but hazards to humans and the environment remain. Science of the Total Environment. 2023;**854**:158814. DOI: 10.1016/j.scitotenv.2022.158814

[2] Bhandari G, Atreya K, Yang X, Fan L, Geissen V. Factors affecting pesticide safety behaviour: The perceptions of Nepalese farmers and retailers. Science of the Total Environment. 2018;**631**:1560–1571. DOI: 10.1016/j.scitotenv.2018.03.144

[3] Zayan SA. Impact of climate change on plant diseases and IPM strategies. Plant Diseases-Current Threats and Management Trends. 2020. DOI: 10.5772/intechopen.87055

[4] Martin P, Ramalanjaona L, Truche C, Ballot R, Carozzi M, Pomeon T. Modelling the spatialisation of pesticide sales to monitor environmental policies in France. Journal of Cleaner Production. 2023;**403**:136880. DOI: 10.1016/j.jclepro.2023.136880

[5] Rani L, Thapa K, Kanojia N, Sharma N, Singh S, Grewal AS, Srivastav AL, Kaushal J. An extensive review on the consequences of chemical pesticides on human health and environment. Journal of Cleaner Production. 2021;**283**:124657. DOI: 10.1016/j.jclepro.2020.124657

[6] Ewald JA, Wheatley CJ, Aebischer NJ, Moreby SJ, Duffield SJ, Crick HQ, Morecroft MB. Influences of extreme weather, climate and pesticide use on invertebrates in cereal fields over 42 years. Global Change Biology. 2015;**21**(11):3931–3950. DOI: 10.1111/gcb.13026

[7] Hallmann CA, Sorg M, Jongejans E, Siepel H, Hofland N, Schwan H, Stenmans W, Müller A, Sumser H, Horren T. More than 75 percent decline over 27 years in total flying insect biomass in protected areas. PLoS One. 2017;**12**(10):e0185809. DOI: 10.1371/journal.pone.0185809

[8] Mehmood Y, Arshad M, Mahmood N, Kachele H, Kong R. Occupational hazards, health costs, and pesticide handling practices among vegetable growers in Pakistan. Environmental Research. 2021;**200**:111340. DOI: 10.1016/j.envres.2021.111340

[9] Rajmohan K, Chandrasekaran R, Varjani S. A review on occurrence of pesticides in environment and current technologies for their remediation and management. Indian Journal of Microbiology. 2020;**60**:125–138. DOI: 10.1007/s12088-019-00841-x

[10] Arya S, Kumar R, Prakash O, Rawat AP. Impact of Insecticides on Soil and Environment and Their Management Strategies. In: Naeem M, Bremont JFJ, Ansari AA, Gill SS, eds. Agrochemicals in Soil and Environment: Impacts and Remediation. Springer: Singapore; 2022. p. 213–230. DOI: 10.1007/978-981-16-9310-6_10

[11] Dhuldhaj UP, Singh R, Singh VK. Pesticide contamination in agro-ecosystems: Toxicity, impacts, and bio-based management strategies. Environmental Science and Pollution

Control Series. 2023;**30**(4):9243–9270. DOI: 10.1007/s11356-022-24381-y

[12] Loos R, Gawlik BM, Locoro G, Rimaviciute E, Contini S, Bidoglio G. EU-wide survey of polar organic persistent pollutants in European river waters. Environmental Pollution. 2009;**157**(2):561–568. DOI: 10.1016/j.envpol.2008.09.020

[13] Brown CD, Van Beinum W. Pesticide transport via sub-surface drains in Europe. Environmental Pollution. 2009;**157**(12):3314–3324. DOI: 10.1016/j.envpol.2009.06.029

[14] Helfrich LA, Weigmann DL, Hipkins PA, Stinson ER. Pesticides and Aquatic Animals: A Guide to Reducing Impacts on Aquatic Systems. VirginiaTech. 2009;24. https://www.pubs.ext.vt.edu/420/420-013/420-013.html

[15] Scholz NL, Fleishman E, Brown L, Werner I, Johnson ML, Brooks ML, Mitchelmore CL, Schlenk D. A perspective on modern pesticides, pelagic fish declines, and unknown ecological resilience in highly managed ecosystems. Bioscience. 2012;**62**(4):428–434. DOI: 10.1525/bio.2012.62.4.13

[16] Hussain S, Siddique T, Saleem M, Arshad M, Khalid A. Impact of pesticides on soil microbial diversity, enzymes, and biochemical reactions. Advances in Agronomy. 2009;**102**:159–200. DOI: 10.1016/S0065-2113(09)01005-0

[17] Rehana Z, Malik A, Ahmad M. Mutagenic activity of the Ganges water with special reference to the pesticide pollution in the river between Kachla to Kannauj (UP), India. Mutation Research/Genetic Toxicology. 1995;**343**(2–3):137–144. DOI: 10.1016/0165-1218(95)90079-9

[18] Agarwal A, Prajapati R, Singh OP, Raza S, Thakur L. Pesticide residue in water—a challenging task in India. Environmental Monitoring and Assessment. 2015;**187**(2):1–21. DOI: 10.1007/s10661-015-4287-y

[19] Dwivedi S, Mishra S, Tripathi RD. Ganga water pollution: A potential health threat to inhabitants of Ganga basin. Environment International. 2018;**117**:327–338. DOI: 10.1016/j.envint.2018.05.015

[20] Ali S, Ullah MI, Sajjad A, Shakeel Q, Hussain A. Environmental and Health Effects of Pesticide Residues, Sustainable Agriculture Reviews 48: Pesticide Occurrence, Analysis and Remediation 2 Analysis. Switzerland AG: Springer Nature; 2021. p. 311–336 DOI: 10.1007/978-3-030-54719-6_8311

[21] Mitra A, Chatterjee C, Mandal FB. Synthetic chemical pesticides and their effects on birds. Research Journal of Environmental Toxicology. 2011;**5**(2):81–96. DOI: 10.3923/rjet.2011.81.96

[22] Ahmad MF, Ali M, Alsayegh ARA, Ahmad S, Alam N, Wahab S, Ali MS, Athar MT. A current novel perspective approach for coronavirus disease-2019 pandemic outbreak. Journal of Advanced Pharmaceutical Technology & Research. 2021;**12**(4):311. DOI: 10.4103/japtr.JAPTR_159_21

[23] Ahmad MF, Wahab S, Ahmad FA, Alam MI, Ather H, Siddiqua A, Ashraf SA, Shaphe MA, Khan MI, Beg RA. A novel perspective approach to explore pros and cons of face mask in prevention the spread of SARS-CoV-2 and other pathogens. Saudi

Pharmaceutical Journal. 2021;**29**(2):121–133. DOI: 10.1016/j.jsps.2020.12.014

[24] Aktar W, Sengupta D, Chowdhury A. Impact of pesticides use in agriculture: Their benefits and hazards. Interdisciplinary Toxicology. 2009;**2**(1):1–12. DOI: 10.2478/v10102-009-0001-7

[25] Thieme T, Heimbach U, Müller A. Chemical Control of Insect Pests and Insecticide Resistance in Oilseed Rape. Biocontrol-based Integrated Management of Oilseed Rape Pests. FAO, Springer; 2010. p. 313–335. https://download.e-bookshelf.de/download/0000/0733/76/L-G-0000073376-0002355210.pdf

[26] Sengül Demirak MS, Canpolat E. Plant-based bioinsecticides for mosquito control: Impact on insecticide resistance and disease transmission. Insects. 2022;**13**(2):162. DOI: 10.3390/insects13020162

[27] Mehmood Y, Arshad M, Mahmood N, Kachele H, Kong R. Occupational hazards, health costs, and pesticide handling practices among vegetable growers in Pakistan. Environmental Research. 2021;**200**:111340. DOI: 10.1016/j.envres.2021.111340

[28] Frangenberg A. Sustainable development of agricultural process and consequences of the implementation of Agenda 21. Integrated crop management as fundamental basis for sustainable production. Pflanzenschutz-Nachrichten Bayer. 2000;**53**(2/3):131–153. hal.science/hal-01186780v1/document

[29] Baker BP, Benbrook CM, Iii EG, Benbrook KL. Pesticide residues in conventional, integrated pest management (IPM)-grown and organic foods: Insights from three US data sets, Food Addit. Contamination. 2002;**19**(5):427–446. DOI: 10.1080/02652030110113799

[30] Nwilene FE, Nwanze K, Youdeowei A. Impact of integrated pest management on food and horticultural crops in Africa. Entomologia Experimentalis Et Applicata. 2008;**128**(3):355–363. DOI: 10.1111/j.1570-7458.2008.00744.x

[31] Chandler D, Davidson G, Grant W, Greaves J, Tatchell G. Microbial biopesticides for integrated crop management: An assessment of environmental and regulatory sustainability. Trends in Food Science & Technology. 2008;**19**(5):275–283. DOI: 10.1016/j.tifs.2007.12.009

[32] Ahmad M, Ahmad FA, Alsayegh AA, Zeyaullah M, AlShahrani AM, Muzammil K, Saati AA, Wahab S, Elbendary EY, Kambal N, Abdelrahman MH, Hussain S. Pesticides impacts on human health and the environment with their mechanisms of action and possible countermeasures. Heliyon. 2024;**10**(7):e29128. DOI: 10.1016/j.heliyon.2024.e29128

[33] Kogan M. Integrated pest management: Historical perspectives and contemporary developments. Annual Review of Entomology. 1998;**43**:243. DOI: 10.1146/annurev.ento.43.1.243

[34] Way M, Van Emden H. Integrated pest management in practice—pathways towards successful application. Crop Protection. 2000;**19**(2):81–103. DOI: 10.1016/S0261-2194(99)00098-8

[35] Palacios Xutuc C. Manual to Train Trainers on Safe and Correct Use of Plant

Protection Products and Integrated Pest Management (IPM). Guatemala City, Guatemala: CropLife Latin America; 2010. croplifeafrica.org/wp-content/uploads/2018/11/TRAINERS.pdf

[36] Van Boxstael S, Habib I, Jacxsens L, De Vocht M, Baert L, Van de Perre E, Rajkovic A, Lopez-Galvez F, Sampers I, Spanoghe P. Food safety issues in fresh produce: Bacterial pathogens, viruses and pesticide residues indicated as major concerns by stakeholders in the fresh produce chain. Food Control. 2013;**32**(1):190–197. DOI: 10.1016/j.foodcont.2012.11.038

[37] Heeb L, Jenner E, Cock MJ. Climate-smart pest management: Building resilience of farms and landscapes to changing pest threats. Journal of Pest Science. 2019;**92**(3):951–969. DOI: 10.1007/s10340-019-01083-y

[38] Fargnoli M, Lombardi M, Puri D, Casorri L, Masciarelli E, Mandi-Rajcevi S, Colosio C. The safe use of pesticides: A risk assessment procedure for the enhancement of occupational health and safety (OHS) management. International Journal of Environmental Research and Public Health. 2019;**16**(3):310. DOI: 10.3390/ijerph16030310

[39] Siriruttanapruk S, Anantagulnathi P. Occupational health and safety situation and research priority in Thailand. Industrial Health. 2004;**42**(2):135–140. DOI: 10.2486/indhealth.42.135

[40] Handford CE, Elliott CT, Campbell K. A review of the global pesticide legislation and the scale of challenge in reaching the global harmonization of food safety standards. Integrated Environmental Assessment and Management. 2015;**11**(4):525–536. DOI: 10.1002/ieam.1635

[41] Zhang W, Jiang F, Ou J. Global pesticide consumption and pollution: With China as a focus. Proceedings of the International Academy of Ecology and Environmental Sciences. 2011;**1**(2):125. DOI: 10.1002/ieam.1635

[42] Nordborg M, Davis J, Cederberg C, Woodhouse A. Freshwater ecotoxicity impacts from pesticide use in animal and vegetable foods produced in Sweden. Science of the Total Environment. 2017;**581**:448–459. DOI: 10.1016/j.scitotenv.2016.12.153

[43] Khoury S. Pesticide land: Brazil's poison market, Revisiting Crimes of the Powerful. Routledge. 2018:174–187. https://www.researchgate.net/profile/David-Whyte-2/publication/328249106_Revisiting_Crimes_of_the_Powerful_Marxism_crime_and_deviance/links/5c59b34645851582c3cff81e/Revisiting-Crimes-of-the-Powerful-Marxism-crime-and-deviance.pdf

[44] Pelaez V, da Silva LR, Araujo EB. Regulation of pesticides: A comparative analysis, Sci. Public Policy. 2013;**40**(5):644–656. DOI: 10.1093/scipol/sct020

[45] Cederberg C. Certification Schemes (RTRS and ProTerra) in Brazilian Soy: Use of Pesticides and Cropping Systems. SIK Institutet for livsmedel och bioteknik; 2013. https://www.diva-portal.org/smash/get/diva2:944188/FULLTEXT01.pdf

[46] Feijao C, d'Angelo C, Flanagan I, Abellan B, Gloinson ER, Smith E, Traon D, Gehrt D, Teare H, Dunkerley F Development of future scenarios for sustainable pesticide use and achievement of pesticide-use and risk-reduction targets announced in the

Farm to Fork and Biodiversity Strategies by 2030. Directorate General for Health and Food Safety March/2022. Brussel: European Commission B-1049. https://food.ec.europa.eu/system/files/2022-06/pesticides_sud_eval_2022_foresight_report.pdf

[47] European Commission. Communication from the Commission to the European Parliament, the Council, the European Economic and Social Committee and the Committee of the Regions. 2020. Brussels, COM: EU Biodiversity Strategy for 2030; 380. https://eur-lex.europa.eu/legal-content/EN/TXT/?uri=celex%3A52020DC0380

[48] European Commission: Directorate-General for Environment, EU Biodiversity Strategy for 2030 – Bringing Nature Back into Our Lives. Publications Office of the European Union; 2021. https://data.europa.eu/DOI/10.2779/677548

[49] Pathak VM, Verma VK, Rawat BS, Kaur B, Babu N, Sharma A, Dewali S, Yadav M, Kumari R, Singh S, Mohapatra A, Pandey V, Rana N, Cunill JM. Current status of pesticide effects on environment, human health and it's eco-friendly management as bioremediation: A comprehensive review. Frontiers in Microbiology. 2022;13. doi:10.3389/fmicb.2022.962619

[50] Sharma S. Cultivating sustainable solutions: Integrated pest management (ipm) for safer and greener agronomy. Corporate Sustainability Management. 2023;**1**(2):103–108. DOI: 10.26480/csmj.02.2023.103.108

[51] Lopez-Antia A, Feliu J, Camarero PR, Ortiz-Santaliestra ME, Mateo R. Risk assessment of pesticide seed treatment for farmland birds using refined field data. Journal of Applied Ecology. 2016;**53**(5):1373–1381. DOI: 10.1111/1365-2664.12668

[52] Kazimierczak R, Średnicka-Tober D, Golba J, Nowacka A, Hołodyńska-Kulas A, Kopczyńska K, Góralska-Walczak R, Gnusowski B. Evaluation of pesticide residues occurrence in random samples of organic fruits and vegetables marketed in Poland. Foods. 2022;**11**(13):1963. DOI: 10.3390/foods11131963

[53] Ssemugabo C, Bradman A, Ssempebwa JC, Guwatudde D. Consumer awareness and health risk perceptions of pesticide residues in fruits and vegetables in Kampala metropolitan area in Uganda. Environmental Health Insights. 2023;17. doi:10.1177/11786302231184751

[54] Ramírez-Santana M, Iglesias-Guerrero J, Castillo-Riquelme M, Scheepers PT. Assessment of health care and economic costs due to episodes of acute pesticide intoxication in workers of rural areas of the Coquimbo region. Value in Health Regional Issues. 2014;**5**:35–39. DOI: 10.1016/j.vhri.2014.07.006

[55] Ren Z, Jiang H. Risk cognition, agricultural cooperatives training, and farmers' pesticide overuse: Evidence from Shandong Province, China. Frontiers in Public Health. 2022;**10**:1032862. DOI: 10.3389/fpubh.2022.1032862

[56] Gangemi S, Gofita E, Costa C, Teodoro M, Briguglio G, Nikitovic D, Tzanakakis G, Tsatsakis AM, Wilks MF, Spandidos DA, Fenga C. Occupational and environmental exposure to pesticides and cytokine pathways in chronic diseases (review). International Journal of Molecular Medicine. 2016;**38**

(4):1012–1020. DOI: 10.3892/ijmm.2016.2728

[57] Bloema BR, Boonstrab TA. The inadequacy of current pesticide regulations for protecting brain health: the case of glyphosate and Parkinson's disease. The Lancet Planetary Health. 2023;**7**(12):e948–e949

[58] Rigal S, Dakos V, Alonso H, et al. Farmland practices are driving bird population decline across Europe. Proceedings of the National Academy of Sciences. 2023;**120**(21):e2216573120. DOI: 10.1073/pnas.2216573120

[59] Finger R, Schneider K, Candel J, Möhring N. Europe needs better pesticide policies to reduce impacts on biodiversity. Food Policy. 2024;**125**:102632. DOI: 10.1016/j.foodpol.2024.102632

[60] Ayilara MS, Adeleke BS, Akinola SA, Fayose CA, Adeyemi UT, Gbadegesin LA, Omole RK, Johnson RM, Uthman QO, Babalola OO. Biopesticides as a promising alternative to synthetic pesticides: A case for microbial pesticides, phytopesticides, and nanobiopesticides. Frontiers in Microbiology. 2023;**14**:1040901. DOI: 10.3389/fmicb.2023.1040901

[61] Whitelaw K. Trends: Innovations and Challenges in Agriculture and Food Systems. [Internet]. 2025. Available from: https://www.dllgroup.com/en/blogs/blogsoverview/2025-trends-innovations-and-challenges-in-agriculture-and-food-systems

[62] LARUS Staff. What is the Internet of Things (IoT)? [Internet]. 2025. https://larus.net/blog/What-is-iot-internet-of-things/?gad_source=1&gad_campaignid=20918745519&gbraid=0AAAAADFRT24LYwmkER8jZxJ4GLqWLJuaf&gclid=CjwKCAiAlfvIBhA6EiwAcErpyZwRSiEqWuiqYfUMu2dLLH-gOtWnDBgHYbHF5i9s8qfIc9yhti4WiBoClaMQAvD_BwE

[63] Etuah S, Adams F, Osei Mensah J, Liu Z, Lan J. Editorial: Increasing resilience and adaptability to climate change of vulnerable groups in agriculture. Frontiers in Sustainable Food Systems. 2024;**8**:1505567. DOI: 10.3389/fsufs.2024.1505567

Chapter 2

Pesticides: Balancing Efficient Pest Control with Environmental and Health Safety

Adeola A. Oso

Abstract

Pesticides have played a pivotal role in modern agriculture, protecting crops and livestock from destructive pests and ensuring global food security. However, their use presents a paradox: while they are effective in safeguarding yields, they pose significant risks to human health, biodiversity, and environmental integrity when misused or overapplied. Despite international bans and restrictions, particularly by the European Union, many developing nations continue to rely heavily on pesticides that are considered hazardous, largely due to weak enforcement systems, limited farmer awareness, and a lack of affordable alternatives. This practice not only heightens health and environmental risks but also leads to the rejection of agricultural produce in international markets, thereby undermining the economic competitiveness of these countries. The consequences of indiscriminate pesticide use are far-reaching. Pesticides contribute to biodiversity decline, contaminate soil and water resources, and disrupt ecological processes critical to sustainable farming. Nontarget organisms, including pollinators, beneficial insects, and aquatic species, are especially vulnerable, creating cascading effects across ecosystems. At the same time, pesticide residues threaten food safety and increase public health burdens, particularly in low- and middle-income countries where regulatory oversight is often weak. Addressing these challenges requires a balanced approach that harmonizes the efficiency of pest control with the protection of environmental and human health. Strategies such as integrated pest management (IPM), biopesticides, and agroecological practices provide viable alternatives to overdependence on synthetic chemicals. Striking this balance is essential not only for sustaining agricultural productivity but also for ensuring biodiversity conservation, safe food systems, and a resilient, sustainable future.

Keywords: biodiversity loss, environmental sustainability, pesticide use, public health, integrated pest management

1. Introduction

Pests remain one of the principal causes of global crop yield losses, accounting for an estimated 20–40% reduction in annual agricultural production [1]. Recent shifts in climate patterns, fluctuating weather conditions, and changing ecosystems have intensified pest activity, stimulating population growth, increasing outbreak frequency, and expanding the geographic distribution of many pest species. Consequently, these developments have contributed to reductions in food security and substantial economic losses [2–4]. Pesticides have long served as indispensable tools in pest management and play a pivotal role in safeguarding global food supplies. According to the Food and Agriculture Organization (FAO) [5], pesticides are defined as any substance or mixture of substances comprising chemical or biological ingredients intended for repelling, destroying, or controlling pests. They are generally classified according to the type of organism they target, including insecticides, fungicides, herbicides, nematicides, acaricides, rodenticides, and bactericides. Historically, these agents have been favored for their low cost, ease of application, rapid action, and high efficacy – helping to stabilize farmers' incomes by minimizing yield losses and improving produce quality. However, decades of intensive and often indiscriminate pesticide use have revealed complex, sometimes subtle, but far-reaching risks to both human health and the environment.

Current global pesticide consumption exceeds four million tons annually [6], encompassing over 1,000 formulations, including chemical, microbial, semichemical, and botanical types. Despite continuous innovations aimed at mitigating adverse effects, even newer selective insecticides have faced regulatory restrictions due to their unintended impacts on pollinators and natural enemies [7, 8], bird populations [9], aquatic life, and overall biodiversity [10]. These cascading ecological effects stem largely from the inherent toxicity and persistence of pesticide compounds [11].

Pesticides and their degradation products have become ubiquitous in the global environment. They have been detected in agricultural soils [12], aquatic ecosystems [13], terrestrial habitats [14], and surface waters [15], often at concentrations exceeding established environmental safety thresholds. Such contamination poses serious ecological and public health challenges, emphasizing the need for environmentally sound chemical management. Achieving Sustainable Development Goals (SDGs) 3.9 and 12.4, which aim to reduce deaths and illnesses from hazardous substances and to promote sustainable production and agriculture by 2030, requires urgent attention to pesticide regulation and safer alternatives.

This paper examines the dual role of pesticides as both essential tools for ensuring food security and significant contributors to environmental degradation. It highlights the need for balanced strategies that reconcile agricultural productivity with ecosystem preservation and human health protection. Sustainable approaches, such as Integrated Pest Management (IPM), the adoption of biopesticides, and the promotion of agroecological practices, offer promising pathways toward reducing dependence on hazardous chemicals while maintaining food production and ecological balance.

2. Global pesticide consumption

The increasing global demand for food, driven by rapid population growth and coupled with a decline in arable land, has made pesticide use indispensable for sustaining agricultural productivity and food security. Pesticides play a vital role in enhancing crop yields and improving food quality, reducing global crop losses due to pests and diseases by approximately 30% [16]. Prior to the advent of synthetic chemical pesticides, pest management relied primarily on rudimentary tools and naturally derived substances, marking a persistent struggle against pest infestations [17]. Although the use of pesticidal substances dates back thousands of years, the widespread adoption of synthetic chemical pesticides began in the 1940s, followed by the rapid development of biopesticides in the last decade. Currently, over 1,000 pesticide formulations are in circulation worldwide, encompassing botanical, chemical, microbial, and semichemical classes. Global pesticide use has increased substantially in recent decades. While the majority are applied in agriculture, an estimated 10–15% are used for noncrop purposes [18]. Globally, agricultural pesticide consumption increased from 2.8 million tons in 2010 to 3.5 million tons in 2022 [19].

The world's four largest agricultural producers – the United States of America, the European Union (EU), China, and Brazil – account for over half of the global agricultural production value [20]. These nations are also among the highest pesticide consumers, using approximately 827 million pounds (EU), 831 million pounds (Brazil), 1.2 billion pounds (USA), and 3.9 billion pounds (China) in 2016 [21].

2.1 Pesticide consumption in Africa

Across the African continent, dependence on pesticides is steadily rising. The main drivers include a shrinking agricultural workforce, rising labor costs, and the need to enhance land productivity amidst competition from imported manufactured goods. Unfortunately, this has created opportunities for individuals with limited technical knowledge to introduce counterfeit or illegally imported pesticides into the market. These counterfeit products, often containing restricted or banned active ingredients, pose serious risks to human health and the environment. The leading countries in pesticide consumption from each region of Africa are shown below (**Table 1**).

2.2 Pesticide consumption in Asia

Trends in pesticide consumption across Asia indicate that pesticides are not only critical inputs for pest management and increased crop yields but are also applied in forestry, public health, and recreational sectors [22]. However, the widespread use of highly hazardous pesticides (HHPs) poses serious threats to human health and the environment [23]. A regional survey in four Asian countries revealed that 27 of 36

S/N	Region	Country	Pesticide use (tons)
1	South	South Africa	26,857
2	North	Egypt	11,352
3	West	Cameroon	7,322
4	Eastern	Ethiopia	4,128

Source: FAO Pesticide Use Data (2022)

Table 1.
Pesticide use in regions of Africa.

commonly used pesticides were highly toxic to humans and ecosystems [24]. Notably, at least seven EU-banned pesticides manufactured by European corporations such as Syngenta, Bayer, and BASF were still in active use, reflecting double standards in pesticide trade and regulatory enforcement.

Key challenges associated with pesticide use in Asia include:

1. The **pesticide treadmill**, characterized by increased pesticide application coupled with declining efficacy, and the unequal distribution of economic gains and environmental burdens;

2. **Adverse effects** on nontarget species, the environment, and human health, including contamination of food and water sources;

3. **Information gaps** and inadequate training among pesticide distributors and users; and

4. **Legislative weaknesses**, poor enforcement of pesticide codes, and frequent regulatory violations.

2.3 Pesticide contamination in the polar region

If any part of the world were expected to be free from pesticide contamination, the Antarctic would be the most likely candidate due to its minimal biological activity. However, organochlorine pesticides (OCPs) have been detected in wildlife near the US Antarctic base at McMurdo Sound, suggesting global dispersion of pollutants via oceanic and atmospheric pathways [25]. A recent study in Prydz Bay, East Antarctica, identified four OCPs and one herbicide (diuron) in surface sediments, with diuron concentrations (below detection limit: 6.57 ng/g) several times higher than those of other OCPs such as p,p′-DDD (≤0.42 ng/g), dichloran (≤0.17 ng/g), β-BHC (≤0.09 ng/g), and HCB (≤0.02 ng/g) [26]. These findings demonstrate the far-reaching impacts of anthropogenic pollution and emphasize the need for global monitoring of pesticide residues, including in remote and pristine ecosystems.

2.4 Pesticide use in Europe

The EU accounts for approximately 13% of global pesticide consumption. Although the rate of increase has been relatively moderate due to stringent regulations, total agricultural pesticide use rose from 402,229 tons in 2010 to 449,038 tons in 2022, representing a 12% increase [27]. As of 2023, the EU had authorized 444 pesticides, banned 954, and placed 43 under further evaluation [EU Pesticides Database, 2023[28]]. Rodríguez-Seijo et al. [29], analyzed 614 pesticides across 188 wheat fields in Europe and detected residues in 151 soils, indicating widespread contamination even under organic and conservation farming systems.

2.5 Pesticide use in the USA

The USA remains the second-largest user of pesticides globally, following China [30]. Pesticides are applied on approximately 900,000 farms and used in about 70 million households. Agricultural pesticide use increased by 10%, from 1.71 million tons in 2021 to 1.89 million tons in 2022 [27]. The rise is largely attributed to the relatively low cost of pesticides compared to alternative pest management methods. Nonetheless, pesticide use in the USA has fluctuated over the past three decades, influenced by factors such as crop prices, weather variations, evolving pesticide regulations, and the adoption of pest-resistant genetically engineered crop varieties [31]. Donley [32] observed that the USA continues to lag behind other major agricultural economies in prohibiting hazardous pesticides. Many active ingredients still widely used, often in quantities reaching hundreds of millions of pounds annually, have already been banned or phased out in the EU, China, and Brazil. Several of these substances have been associated with acute poisoning cases and environmental degradation in the USA, underscoring the need for stronger federal pesticide reform and risk management.

3. Environmental impact of pesticides

While pesticides have made invaluable contributions to global agricultural productivity, their misuse and overreliance have generated profound environmental and public health concerns. It is estimated that less than 15% of applied pesticides reach their intended targets, while the remaining proportion disperses into surrounding soil, air, and water [33]. This off-target deposition contributes to soil and water contamination, the poisoning of nontarget organisms, loss of biodiversity, and the development of resistance in pest populations. As climate change intensifies, increased rainfall, surface runoff, and glacial ice melt exacerbate pesticide mobility and persistence. Legacy pesticide chemicals, banned or no longer in production, are being released from melting ice, contributing to the reintroduction of toxic compounds into aquatic systems [34]. The lack of region-specific monitoring programs further compounds the risk of these chemicals accumulating and interacting synergistically in aquatic environments, leading to chronic water pollution.

3.1 Pesticides in water systems

Pesticides can infiltrate drinking water sources through multiple pathways. One of the most significant is agricultural runoff. Large quantities of pesticides applied to crops and soil are mobilized by rainfall or irrigation water, percolating through soil pores and fractures into groundwater-bearing aquifers. The extent of contamination depends on several factors, including the pesticide's solubility and mobility in soil, the chemical formulation (active ingredients and additives), degradation rate, soil temperature, microbial activity, and irrigation management practices. Overirrigation, for instance, increases the likelihood of surface runoff and leaching, thereby enhancing the transport of pesticides into both surface and groundwater systems. Beyond agricultural use, improper handling and disposal of pesticides from manufacturing plants, mixing-and-loading facilities, and wastewater discharge sites also contribute to contamination. Spillage, waste mismanagement, and leaching from landfills or sewage treatment plants introduce pesticide residues into nearby water bodies. In addition, wind drift during aerial or tractor-based spraying can transport pesticides far beyond target sites, depositing them into rivers, lakes, and other open water sources [35].

Pesticides have been detected globally in rivers, lakes, and oceans, often at sublethal concentrations that nonetheless exert significant biological effects. The European Environmental Agency reports (2013–2022) revealed that one or more pesticides were found above quality thresholds at different monitoring sites. Studies by the United States Geological Survey reported pesticides in nearly every stream and over 90% of sampled wells in the USA. In the Lake Victoria Basin in Kenya, fish poisoning from the pesticide endosulfan was reported [36]. The prevalence of the organophosphate pesticide chlorpyrifos has been documented in water and sediment samples in Bangladesh [37]. Runoff of diuron and atrazine (herbicides) from sugarcane fields in Australia has polluted waterways that drain into the Great Barrier Reef [38]. Long before reaching lethal doses, pesticide residues can disrupt endocrine function, impair neurological processes, and alter behavior in aquatic organisms such as fish [39]. In a recent study [40], it was demonstrated that exposure to chemical pesticides reduces sociability and territorial behavior in fish, resulting in diminished group interactions and fewer mating attempts. During heavy rainfall, pesticides from agricultural fields are carried via runoff into roadside drains and streams, where they eventually accumulate in aquatic ecosystems. Fish absorb these compounds through their gills or by consuming contaminated prey, leading, in severe cases, to large-scale fish mortality.

3.2 Pesticide residues and soil persistence

Pesticide residues are now widespread in agricultural soils across Europe and beyond. Rodríguez-Seijo et al. [29], investigated 614 pesticides in 188 wheat fields across eight European countries, comprising both conventional (n = 93) and organic (n = 95) systems. Their findings revealed significant regional variation. The continental region (Germany) exhibited the highest residue concentrations (mean: 0.46 mg/kg) and diversity (average of 13.5 pesticides per plot), followed by the Atlantic region (Denmark and Belgium). The Pannonian region (Hungary and Serbia) showed the lowest levels (mean: 0.02 mg/kg). A particularly alarming discovery was the detection of pesticide residues in organic fields, where 35 different compounds were identified, only one of which (Spinosad) is permitted in organic farming. This finding suggests both the long-term persistence of legacy pesticides and the potential for cross-contamination between adjacent fields. Moreover, 31 of the detected compounds were banned at the time of sampling, demonstrating that certain pesticides remain detectable more than four decades after their prohibition. These results underscore the high persistence and environmental mobility of specific active ingredients, as well as the long-lasting legacy of historical pesticide use on soil and ecosystem health.

4. Health impacts of pesticides

Populations living in proximity to agricultural lands face increased exposure to harmful pesticide residues. Among these, glyphosate, one of the most widely used herbicides globally, has drawn significant public health attention [41]. Numerous studies have linked glyphosate exposure to various health outcomes, including non-Hodgkin lymphoma and an elevated risk of preterm birth [42]. Curl et al. [43], monitored glyphosate levels in pregnant women over a 10-month period and found that those residing within 500 meters of agricultural fields had significantly higher urinary glyphosate concentrations compared to individuals living farther away. Elevated levels were recorded only during active spraying periods, suggesting direct exposure from agricultural application. Similarly, Hyland et al. [44], observed that consuming organic foods – produced without synthetic pesticides – reduced glyphosate levels in women residing far from agricultural areas, but not in those living nearby, reinforcing the influence of environmental proximity. Emerging evidence further implicates glyphosate exposure during pregnancy in adverse reproductive outcomes, including preterm birth, shortened gestation, and impaired fetal growth [45]. Beyond glyphosate, pesticide residues also infiltrate the food chain through contaminated crops and water sources [46], increasing the risk of multiple chronic diseases. Reported health impacts include neurodegenerative, cardiovascular [47], endocrine, respiratory, renal, and reproductive disorders, as well as various forms of cancer.

A decade-long investigation by the US Geological Survey's National Water-Quality Assessment (NAWQA) program revealed the widespread presence of pesticides in both surface and groundwater across rural and urban areas of the USA [34]. OCPs and their degradation products, such as dichlorodiphenyltrichloroethane (DDT), were frequently detected, reflecting their long-term persistence in soil, sediment, and biota. Similarly, a study by 48 documented the presence of neonicotinoids in US water systems, with at least one compound detected in 53% of all samples and imidacloprid appearing most frequently (37%). The increasing concentration of such chemicals in aquatic environments poses serious risks not only to aquatic organisms but also to terrestrial species and humans relying on these water sources.

Numerous epidemiological studies have established associations between pesticide exposure and neurological disorders. Yan et al. [49], reported that chronic exposure to pesticides is linked to an elevated risk of Alzheimer's disease, while Ahmed et al. [50], found a strong correlation between pesticide exposure and Parkinson's disease, potentially mediated by genetic alterations in dopaminergic pathways. In a cohort of over 30,000 female spouses of pesticide applicators, Lerro et al. [51] demonstrated that higher organophosphate exposure significantly increased the risk of hormone-related cancers, including breast, thyroid, and ovarian cancer. Children are particularly vulnerable to pesticide toxicity due to their developing organs and metabolic systems. Accidental or chronic exposure to elevated pesticide levels has been associated with autism spectrum disorder, leukemia, lymphoma, and attention deficit hyperactivity disorder (ADHD). Even low-level exposure has been linked to adverse neurological and behavioral outcomes [52]. A study by Sheahan et al. [53], across four African countries – Ethiopia, Nigeria, Tanzania, and Uganda – examined the relationship between pesticide use, agricultural productivity, and human health indicators. Although pesticide application correlated with higher crop output values, it also imposed considerable human health costs, including illness-related productivity loss and reduced labor supply. These findings highlight the complex trade-off between the economic benefits of pesticide use and its profound implications for public health.

5. Pesticide impact on biodiversity

Arthropods, including bees, beetles, butterflies, millipedes, spiders, and numerous others, constitute more than 80% of all known animal species. Though often perceived merely as pests, these organisms are fundamental to ecosystem stability: they pollinate crops, enrich and aerate soils, regulate pest populations, and sustain the ecological networks upon which human survival depends. However, industrial agriculture and the indiscriminate use of pesticides have driven alarming declines in their populations [54]. The history of pesticide use is replete with cases of environmental devastation, from DDT and parathion to paraquat, chlorpyrifos, neonicotinoids, and chlorothalonil. Many compounds once considered safe for humans and wildlife have since been banned after evidence of their far-reaching ecological toxicity emerged [55]. These patterns underscore an ongoing biodiversity crisis. Modern insect populations are exposed not to single chemicals but to complex "chemical cocktails" of insecticides, herbicides, and fungicides, often interacting synergistically in the environment [56].

Traditional pesticide risk assessments focus largely on acute lethality, neglecting sublethal and chronic effects that disrupt essential biological functions. Recent research from the European Molecular Biology Laboratory highlights this limitation: screening over 1,000 agrochemicals using a *Drosophila* model, Gandara et al. [57] found that many compounds induced behavioral and developmental alterations at sublethal concentrations. When agrochemicals were combined at field-realistic doses, the researchers observed widespread disruptions in larval development, behavior, and reproduction. These findings indicate that even minimal, chronic exposure can significantly impair insect populations. Moreover, the toxicity of several insecticides, including lindane, increased dramatically under elevated temperatures, rising from 0% mortality at 25°C to 79% at 29°C, underscoring the compounding effects of climate change on pesticide toxicity. Globally, biodiversity indicators mirror this crisis. The Living Planet Report [58] estimated that wild vertebrate populations have declined by 73% since 1970. Terrestrial vertebrate body sizes and ranges have contracted by approximately one-third, with many mammal species experiencing range reductions exceeding 80% over the last century [59]. Amphibians are among the most imperiled taxa, with nearly half of all species currently at risk of extinction [60]. Insects, though less comprehensively monitored, show equally disturbing trends: global assessments suggest an average decline of 1–2% per year [61].

Aquatic ecosystems are similarly vulnerable. Pesticide contamination in waterways exerts a profound "bottom-up" effect by decimating aquatic invertebrates such as benthic organisms and cladocerans. These organisms play essential roles in nutrient cycling, detritus decomposition, and as prey for higher trophic levels. Many benthic invertebrates are larval stages of terrestrial insects, serving as critical indicators of water quality and biodiversity [34]. The neurotoxic mechanisms of most insecticides do not distinguish between aquatic and terrestrial taxa, leading to cascading disruptions across aquatic food webs. Collectively, these findings highlight the urgent need to reassess pesticide risk frameworks to incorporate long-term, cumulative, and ecosystem-level effects. Without substantial reform in pesticide management and agricultural practices, continued biodiversity loss may irreversibly compromise ecosystem resilience, agricultural productivity, and global food security.

6. Sustainable and integrated pest management (IPM) approaches

The growing evidence of adverse environmental and health consequences associated with pesticide use has prompted global calls for safer and more sustainable pest control strategies. IPM represents a holistic approach that combines biological, cultural, mechanical, physical, and chemical control methods to maintain pest populations below economically damaging thresholds while minimizing risks to human health and the environment [62]. The IPM concept emphasizes prevention, monitoring, and the use of control measures that are both ecologically and economically sustainable. Pests and diseases pose a severe threat to food security, trade, and livelihoods globally [63]. Each year, up to 40% of crops are lost to plant pests and diseases [1]. Conventional pest control practices, characterized by the heavy and often indiscriminate use of chemical pesticides, have led to multiple ecological and societal challenges, including the emergence of pesticide resistance, the disruption of beneficial arthropod communities, contamination of soil and water, and the exposure of farm workers and consumers to hazardous residues [64]. These challenges underscore the urgent need for ecologically sound pest management systems.

6.1 Principles and components of integrated pest management (IPM)

A central tenet of IPM is ecological balance. Rather than relying solely on chemical control, IPM integrates natural regulatory mechanisms that enhance ecosystem resilience. Biological control, using natural enemies such as predators, parasitoids, and pathogens, plays a pivotal role. Beneficial insects like lady beetles, lacewings, and parasitoid wasps naturally suppress aphid and caterpillar populations, reducing the need for synthetic insecticides. The conservation of these beneficial organisms through habitat diversification, reduced pesticide exposure, and ecological engineering enhances ecosystem stability and pest regulation capacity.

Cultural practices such as crop rotation, intercropping, and the use of pest-resistant varieties also disrupt pest life cycles and limit population build-up [65]. Crop rotation, for instance, involves sequentially planting different crops in the same field [66]. This practice disrupts pest host cycles, suppresses disease inoculum, and improves soil fertility, contributing to yield enhancement. The efficacy of crop rotation is linked to the spatial and temporal separation of host crops, the incorporation of nonhost or trap crops, and the promotion of biodiversity that supports natural enemies [67].

The use of resistant varieties represents another essential IPM strategy. By exploiting genetic diversity, farmers can reduce pest-induced damage and limit chemical dependency [68]. Resistance mechanisms fall into three categories: *antixenosis* (deterring pest feeding or oviposition), *antibiosis* (producing compounds that impair pest development or survival), and *tolerance* (the ability to withstand or recover from damage without yield loss) [69, 70]. Integrating resistant varieties with other management strategies strengthens system resilience and sustainability.

Cultural and mechanical practices, including field sanitation, timely planting, proper tillage, pruning, and the use of physical barriers, further create environments less conducive to pest infestations. Maintaining healthy soils through organic amendments and cover cropping supports beneficial soil microorganisms that naturally suppress pest populations.

Monitoring and decision-making are also central to IPM. Farmers regularly assess pest population dynamics using traps, field scouting, and predictive tools to determine the economic threshold for intervention. This data-driven approach ensures that control measures are applied only when necessary, minimizing chemical use. Empirical studies show that regular pest monitoring can reduce pesticide use by up to 50% without compromising crop yields [71]. Collectively, these integrated strategies minimize pest damage, conserve biodiversity, and safeguard ecosystem services, thereby strengthening the long-term stability and productivity of agricultural systems [70].

6.2 Biopesticides and botanical alternatives

Biopesticides and botanical formulations are increasingly recognized as key components of sustainable pest management. Derived from natural materials such as plants, bacteria, fungi, and minerals, these products exhibit low toxicity, target specificity, and rapid biodegradability. They are, therefore, less harmful to nontarget organisms and the environment compared to conventional pesticides [72]. Prominent examples include *Bacillus thuringiensis* (Bt) for the control of lepidopteran larvae, neem (*Azadirachta indica*)-based formulations effective against insects [73] and fungi, and entomopathogenic fungi such as *Beauveria bassiana* and *Metarhizium anisopliae* for pest suppression [74]. The use of such bioinsecticides and botanical extracts helps to reduce soil and water contamination, preserve biodiversity, and enhance the sustainability of pest control programs. Although the commercial adoption of biopesticides remains limited compared to conventional chemicals, advances in formulation technology, improved field efficacy, and supportive regulatory frameworks are expanding their market potential. Their integration within IPM programs aligns with the principles of agroecology and climate-smart agriculture.

6.3 Technological innovations in IPM implementation

Modern precision agriculture technologies further enhance IPM implementation by enabling real-time monitoring, targeted pesticide applications, and data-driven decision-making. Tools such as drones, satellite imagery, geographic information systems (GIS), and digital pest surveillance platforms can detect early pest outbreaks, allowing for localized interventions and substantial reductions in chemical inputs [75]. These innovations promote efficient resource use, reduce costs, and minimize environmental contamination while supporting biodiversity conservation. The adoption of these innovations requires substantial investment, education, and infrastructural development, particularly in developing regions. In addition, a balance between utilizing technology to reduce pests and disease, along with not forgetting the benefits of traditional knowledge, needs to be maintained.

6.4 Toward sustainable pest management

Achieving sustainable pest management demands a paradigm shift from dependence on chemical control to ecosystem-based approaches that value biodiversity, soil health, and ecological integrity. IPM, when combined with agroecological principles and green innovation, offers a realistic pathway to safeguard global food security while minimizing environmental and health burdens associated with pesticide use. The transition toward IPM not only enhances agricultural sustainability but also aligns with the broader objectives of the United Nations' SDGs, particularly those related to zero hunger, good health, and life on land.

7. Policy and regulatory framework

Policy and institutional support are essential for scaling up IPM adoption globally. Governments and regulatory agencies must strengthen pesticide legislation, streamline the registration of safer biopesticides, and phase out hazardous chemicals. Extension services and farmer education programs are vital for promoting awareness and practical adoption of IPM principles. Furthermore, research collaboration among scientists, policymakers, and farmers ensures that pest management strategies remain economically viable, socially acceptable, and environmentally sound [62]. However, the various regulatory systems for pesticide approval currently in operation around the world are often crude and flawed [55, 76]. This inadequacy is deeply concerning, as these systems are intended to protect both people and the environment from harm. The EU regulatory framework for pesticides is often regarded as the most rigorous globally; yet, it has repeatedly approved substances later found to be harmful to humans and wildlife, leading to subsequent bans. In 2019, the PEST Committee of the European Parliament concluded, after nine months of investigations, hearings, and commissioned studies, that the current pesticide authorization system was failing to achieve its intended protective goals. The Committee's report, endorsed by the European Parliament [77], emphasized the urgent need for reform.

Subsequent analysis reported in [78] revealed limited progress in implementing the PEST recommendations: only 15% had been sufficiently implemented, 28% were partially or insufficiently implemented, and 57% had not been implemented at all, with some areas showing further deterioration. Major weaknesses in the EU and UK pesticide regulatory systems include:

i. Regulatory testing focuses on the active substance, while the actual commercial formulations containing additional synergistic ingredients are rarely evaluated, despite their potential to amplify toxicity.

ii. Toxicity tests for wildlife are often conducted in-house by pesticide manufacturers, and the data are withheld from the public under claims of commercial confidentiality.

iii. Testing protocols generally assess only short-term (often 48-hour) exposure effects in healthy test species, such as bees, beetles, or fish, disregarding chronic, cumulative, or sublethal impacts.

iv. Standard tests expose organisms to single pesticides, even though real-world exposure involves complex mixtures that may interact synergistically to increase toxicity.

If EU pesticide regulations were fully enforced and risk assessment methods revised to be scientifically rigorous, objective, and transparent, many substances currently deemed safe would likely be found to pose unacceptable risks to human health and the environment and would require prohibition or restriction.

At the global level, several international bodies, including the FAO, the World Health Organization (WHO), and the EU, have established conventions and guidelines to minimize pesticide-related risks. Notably, the Rotterdam Convention (1998) regulates international trade in hazardous chemicals, while the Stockholm Convention (2001) governs persistent organic pollutants (POPs). The EU has enacted stringent bans and restrictions on several active ingredients, such as chlorpyrifos, paraquat, and neonicotinoids, due to their proven adverse impacts on biodiversity, pollinators, and human health [79]. Similarly, the FAO/WHO International Code of Conduct on Pesticide Management sets global standards for pesticide registration, distribution, use, and disposal, emphasizing the transition to safer alternatives through IPM adoption.

Despite these international commitments, weak enforcement and institutional constraints persist in many developing countries. Challenges include inadequate policy implementation, limited laboratory infrastructure for residue analysis, poor monitoring systems, and weak regulatory oversight [80]. Furthermore, porous borders, unregulated informal markets, and insufficient data on pesticide imports facilitate the continued use of banned or restricted substances. Consequently, hazardous active ingredients prohibited in the EU remain widely available and applied in low- and middle-income countries, posing significant threats to human health, biodiversity, and water quality.

National governments must therefore play a central role in closing these regulatory gaps through stronger legislation, continuous product reevaluation, and enhanced enforcement mechanisms. Establishing national pesticide control boards, residue monitoring laboratories, and data-driven surveillance systems can improve transparency and accountability across the pesticide supply chain. Harmonizing national policies with international conventions also fosters safer trade practices and compliance with global safety standards.

Equally important is the education and capacity building of farmers and extension workers. Many smallholder farmers lack adequate training in pesticide safety, risk awareness, and alternative pest management approaches. Strengthening extension services, farmer field schools, and community-based IPM programs can enhance awareness and promote the adoption of sustainable pest control methods. Partnerships among governments, research institutions, and nongovernmental organizations (NGOs) are critical to ensure that policy reforms lead to behavioral change and practical impact. A robust and transparent policy and regulatory framework, supported by international cooperation, national enforcement, and farmer education, is fundamental to achieving sustainable pesticide management. Aligning national policies with global standards while empowering farmers with knowledge and institutional support will be key to reducing pesticide-related hazards and advancing public health and environmental sustainability.

8. Trade-offs and feasibility

The extensive reliance on pesticides presents a profound dilemma for global agriculture – how to sustain high productivity while protecting environmental integrity and human health. The persistent trade-offs between efficiency and sustainability, profit and well-being, and innovation and equity remain central to the ongoing debate over agricultural modernization. Addressing these challenges requires a nuanced evaluation of both the economic and ecological costs of pesticide use, alongside the feasibility of transitioning to more sustainable pest management approaches. Pesticides have long been essential tools for protecting crops, reducing postharvest losses, and stabilizing food supply chains. Their use has contributed significantly to agricultural profitability and global food security. However, these short-term benefits come with substantial environmental and health consequences. Persistent residues contaminate soils and water bodies, disrupt aquatic ecosystems, and affect nontarget species such as pollinators and natural predators. These disturbances reduce biodiversity, weaken ecosystem services, and ultimately foster pesticide resistance – creating a cycle of chemical dependency and ecological imbalance [11, 78]. From an economic standpoint, pesticides are valued for their affordability, ease of application, and proven efficacy – factors that make them indispensable in intensive farming systems. Yet, these apparent advantages conceal broader externalities. The economic burden associated with healthcare costs, loss of labor productivity, and environmental remediation often surpasses the short-term financial gains achieved through pesticide use. In many developing nations, weak regulatory oversight, poor enforcement mechanisms, and inadequate farmer training exacerbate these challenges, resulting in frequent misuse, unsafe exposure, and environmental contamination [80]. Emerging alternatives, including genetically engineered crops, biopesticides, and precision agriculture technologies, offer promising opportunities to reconcile productivity with sustainability. Biopesticides derived from natural organisms, for instance, minimize nontarget toxicity and biodegrade more rapidly, while precision monitoring tools enable targeted and efficient pesticide applications. Similarly, the adoption of genetically modified pest-resistant crops has demonstrated potential in reducing chemical reliance. However, access to these innovations remains uneven. Low-income countries often face barriers such as high implementation costs, limited technical expertise, inadequate infrastructure, and insufficient institutional support. Consequently, these regions continue to depend on older, inexpensive, and frequently more hazardous chemical formulations, thereby widening the global divide in agricultural sustainability [18]. Ultimately, the trade-offs between productivity and sustainability cannot be entirely eliminated, but they can be effectively mitigated through context-specific, science-driven strategies. Strengthening pesticide regulation and enforcement, promoting IPM, and investing in research on low-toxicity alternatives are critical policy priorities. Furthermore, building institutional capacity, enhancing farmer education, and ensuring equitable access to safe technologies are essential for achieving sustainable agricultural transformation. By aligning innovation with environmental stewardship and public health protection, the global community can move toward a more balanced and feasible framework for pest management – one that preserves productivity without compromising ecological and human well-being.

9. Balancing efficiency and safety

The agricultural sector faces a critical challenge in balancing the demand for increased productivity with the imperative of ensuring environmental sustainability and public health. As the global population continues to rise, projected to reach nearly 10 billion by 2050, the pressure to maximize crop yields has intensified [1]. Chemical pesticides have played a central role in achieving these yield gains by reducing pest-related losses, which account for an estimated 20–40% of global crop yield reductions annually [81]. However, this dependence has generated significant trade-offs, including environmental pollution, pest resistance, biodiversity loss, and health hazards to farm workers and consumers [64]. The challenge, therefore, lies in balancing the efficiency of pest control with the safety of the environment and human health – a balance that underpins sustainable agricultural development.

The dilemma of productivity versus sustainability reflects the broader tension between short-term agricultural gains and long-term ecological stability. In many developing countries, the drive to increase productivity often leads to the indiscriminate and excessive use of pesticides, sometimes including chemicals that have been banned in the EU or restricted under international conventions [80]. This trend is exacerbated by weak regulatory frameworks, limited farmer education, and inadequate access to safer alternatives. While these chemicals can deliver immediate control of pest outbreaks, they can also result in pesticide residues in food, contamination of water sources, and soil degradation, ultimately undermining future agricultural potential.

Principles for responsible pesticide use are, therefore, essential for minimizing risks while maintaining productivity. These principles include adopting the "as low as reasonably achievable" (ALARA) principle for dosage and frequency of pesticide application, following label instructions, observing preharvest intervals, and ensuring the safe storage and disposal of pesticide containers [82]. Precision agriculture technologies, such as drone-based spraying and sensor-guided application systems, can enhance pesticide efficiency by ensuring accurate targeting and minimizing waste. Moreover, selecting selective or low-toxicity pesticides can protect beneficial insects, pollinators, and soil microorganisms that contribute to natural pest suppression and overall ecosystem health.

Sustainable pest management approaches, particularly IPM, offer long-term solutions to this dilemma. IPM integrates biological, cultural, mechanical, and chemical control methods, emphasizing ecological balance rather than pest eradication [70]. Studies have shown that IPM implementation can reduce pesticide use by up to 50% without compromising yields, while also enhancing soil quality and biodiversity [71]. Similarly, the adoption of biopesticides and botanical extracts, such as neem-based formulations, has gained traction in countries like India, Kenya, and Nigeria, offering effective, biodegradable, and residue-free pest control alternatives [63, 73]. These innovations contribute to sustainable agriculture by reducing the chemical footprint and preserving ecosystem services.

The long-term benefits of sustainable approaches extend beyond environmental gains to include economic and social advantages. Farmers benefit from reduced input costs, improved soil fertility, and increased access to high-value markets that demand pesticide residue-free produce. Consumers benefit from safer food, while communities experience reduced exposure to toxic chemicals and cleaner environments. At the national level, countries that adopt sustainable pest management practices strengthen their compliance with international standards, such as the FAO/WHO International Code of Conduct on Pesticide Management and the SDGs, particularly SDG 2 (Zero Hunger) and SDG 15 (Life on Land).

Ultimately, balanced pest control supports food security, health, and environmental integrity. A well-regulated and knowledge-driven pest management system ensures that agricultural productivity is sustained without compromising future generations' ability to meet their own needs. Policymakers, researchers, and farmers must collaborate to strengthen regulatory frameworks, promote capacity building, and support innovation in safer pesticide technologies. Balancing efficiency and safety in pest management is not merely a technical goal but a moral and developmental imperative – one that aligns with the global vision for resilient, safe, and sustainable food systems.

10. Recommendations and conclusion

Pesticides play a dual role in modern agriculture, serving as indispensable tools for enhancing crop yields and protecting food supplies, yet simultaneously posing substantial risks to human health and the environment. While their use has contributed to significant productivity gains, the persistence of chemical residues in soil, water, and food chains underscores the urgent need for more sustainable pest management approaches. The challenge lies in balancing the efficiency of pesticide use with the imperative of safeguarding ecological integrity and public well-being.

Achieving this balance requires integrated and adaptive solutions that combine scientific innovation, sound policy frameworks, and local farming knowledge. The promotion of IPM stands out as the most viable approach, harmonizing biological, cultural, mechanical, and chemical control methods. By prioritizing prevention, monitoring, and the judicious use of pesticides as a last resort, IPM reduces environmental contamination, delays pest resistance, and supports long-term agricultural sustainability. Governments should strengthen pesticide legislation, phase out highly hazardous compounds, and invest in research on safer alternatives such as biopesticides and natural pest enemies. Building farmers' capacity through extension services, public education, and participatory training is equally vital to promote behavioral change and ensure effective implementation of IPM practices. Multisectoral collaboration among policymakers, researchers, agro-industries, and local communities will foster context-specific solutions that align with national and global sustainability goals.

In conclusion, the future of pest management lies in fostering a resilient agricultural system that optimizes productivity without compromising environmental and human health. Embracing innovation, reinforcing regulation, and valuing indigenous knowledge will be key drivers of this transformation. A collective commitment to responsible pesticide use will ensure that agriculture remains productive, safe, and sustainable for generations to come.

Acknowledgments

The authors acknowledge the use of ChatGPT for language editing in the chapter.

Author details

Adeola A. Oso
Department of Crop, Horticulture, and Landscape Design, Ekiti State University, Ado-Ekiti, Nigeria

*Address all correspondence to: adeola.oso@eksu.edu.ng

References

[1] FAO. Pests and pesticide management. 2025. https://www.fao.org/pest-and-pesticide-management/ipm/integrated-pest-management/en/ [Accessed: 2025-September-25]

[2] Andrew NR, Hill SJ, Binns M, et al. Assessing insect responses to climate change: What are we testing for? Where should we be heading? Peer Journal. 2013;**1**:e11

[3] Lehmann P, Ammunnét T, Barton M, et al. Complex responses of global pests to climate warming. Frontiers in Ecology and the Environment. 2020;**18**(3):141–150. DOI: 10.1002/fee.2160

[4] Thackeray SJ, Henrys PA, Hemming D, et al. Phenological sensitivity to climate across taxa and trophic levels. Nature. 2016;**535**:241–245

[5] Food and Agriculture Organization of the United Nations. Q&A on Pests and Pesticide Management. FAO. 2021, May 12. Retrieved from https://www.fao.org/newsroom/detail/Q-A-on-Pests-and-Pesticide-Management/fr [Accessed: 2025-September-25]

[6] FAO. FAOSTAT Statistical Database: Pesticides Use [Data Set]. 2019. Food and Agriculture Organization of the United Nations. Retrieved from https://www.fao.org/faostat/en/#data/RP [Accessed: 2025-September-25]

[7] Roubos CR, Rodriguez-Saona C, Isaacs R. Mitigating the effects of insecticides on arthropod biological control at field and landscape scales. Biological Control. 2014;**75**:28–38. DOI: 10.1016/j.biocontrol.2014.01.006

[8] Rundlof M, Andersson GKS, Bommarco R, Fries I, Hederstrom V, Herbertsson L, et al. Seed coating with a neonicotinoid insecticide negatively affects wild bees. Nature. 2015;**521**:77–80

[9] Hallmann CA, Foppen RPB, van Turnout CAM, de Kroon H, Jongejans E. Declines in insectivorous birds are associated with high neonicotinoid concentrations. Nature. 2014;**511**:341–343. DOI: 10.1038/nature

[10] Sánchez-Bayo F, Wyckhuys K. Worldwide decline of the entomofauna: A review of its drivers. Biological Conservation. 2019;**232**:8–27. DOI: 10.1016/j.biocon.2019.01.020

[11] World Health Organization and Food and Agriculture Organization. Report of the 8th FAO/WHO Joint Meeting on Pesticide Management (JMPM) and the 10th Session of the FAO Panel of Experts on Pesticide Management: 14–17 October 2014. Rome: Italy. WHO; 2014

[12] Silva V, Mol HG, Zomer P, Tienstra M, Ritsema CJ, Geissen V. Pesticide residues in European agricultural soils–A hidden reality unfolded. The Science of the Total Environment. 2019;**653**:6531532–1545. DOI: 10.1016/j.scitotenv.2018.10.441

[13] Socorro J, Durand A, Temime-Roussel B, Gligorovski S, Wortham H, Quivet E. The persistence of pesticides in atmospheric particulate phase: An emerging air quality issue. Scientific Reports. 2016;**6**:1–7. DOI: 10.1038/srep33456

[14] Zhang WJ, van der Werf W, Pang Y. A simulation model for vegetable-

insect pest-insect nucleopolyhedroviral epidemic system. Journal of Environmental Entomology. 2011;**33**:283–301

[15] Wagh V, Mukate S, Muley A, Kadam A, Panaskar D, Varade A. Study of groundwater contamination and drinking suitability in basaltic terrain of Maharashtra, India through PIG and multivariate statistical techniques. Journal of Water Supply: Research and Technology-Aqua. 2020;**69**:398–414. DOI: 10.2166/aqua.2020.108

[16] Walter JM, Luciana PV, Ariane FM, Júlio C, Kim KV, Maria CM, Thamarys S, Roberta P, Rocky KL, Vanete T, Carlos RS. Bioremediation strategies against pesticides: An overview of current knowledge and innovations. Chemosphere. 2024;**364**. DOI: 10.1016/j.chemosphere.2024.142867

[17] Isman MB.Challenges of pest management in the twenty first century: New tools and strategies to combat old and new foes alike. Frontiers Agronomy. 2023;**1**(2). DOI: 10.3389/fagro.2019.00002

[18] Agrow. Pesticides Market Overview and Non-crop Pesticide Value Share [Report]. United Nations Environment Programme. Pesticide's pollution and diagnostics: A global assessment. 2019

[19] Food and Agriculture Organization of the United Nations. Pesticides Use and Trade, 1990–2022: FAOSTAT Pesticides Use Database. [Analytical brief]. FAO; 2022. https://www.fao.org/statistics/highlights/archive/highlights-detail/pesticides-use-and-trade-1990-2022/en [Accessed: 2025-September-25]

[20] Food and Agriculture Organization of the United Nations FAOSTAT Database Compare Data 2016. https://www.fao.org.faostat/en/#compare [Accessed: 2025-September-25]

[21] European Commission. Eurostat Database. Pesticide Sales. 2016. https://appsso.eurostat.ec.europa.eu/nui/show.do? [Accessed: 2025-September-25]

[22] Soerjani M. Current Trends in Pesticide Usage in Some Asian Countries: Environmental Implications and Research Needs. Vienna, Austria: IAEA; 1988. 219–234

[23] Dhoj GCY, Palikhe BR, Gu B, Beatrice G. Status of highly hazardous pesticides and their mitigation measures in Asia. Advances in Entomology. 2022;**10**:14–33. DOI: 10.4236/ae.2022.101002

[24] PAN. New report: Majority of pesticides used in farming communities of four Asian countries are highly hazardous or banned in EU. 2022. https://panap.net/2023/02/new-report-majority-of-pesticides-used-in-farming-communities-in-4-asian-countries-are-highly-hazardous-or-banned-in-eu/#_ednref3 [Accessed: 2025-September-25]

[25] Tatton JOG, Ruzicka JHA. Organochlorine pesticides in Antarctica. Nature. 1967;**215**(5099):346–348. DOI: 10.1038/215346a0

[26] John PM, Sanjeevan VN, Narayanapillai AK, Gopinath A. Distribution of organochlorine pesticides and diuron (urea pesticide) in the sediments of Prydz Bay, East Antarctica. Arabian Journal of Geosciences. 2025;**18**:102. DOI: 10.1007/s12517-025-12247-9

[27] Food and Agriculture Organization of the United Nations FAOSTAT Database Compare Data 2024. https://

www.fao.org/statistics/highlights-archive/highlights-detail/pesticides-use-and-trade-1990-2022/en [Accessed: 2025-September-25]

[28] European Commission 2023. EU Pesticides Database. https://food.ec.europa.eu/plants/pesticides/eu-pesticides-database_en [Accessed: 2025-September-29] (Luxembourg)

[29] Rodríguez-Seijo A, Pérez-Rodríguez P, Arias-Estévez M, Gómez-Armesto A, Conde-Cid M, Santás-Miguel V, Campillo-Cora C, Ollio I, Lloret E, Martínez-Martínez S, Zornoza R, Waeyenberge L, Schrader S, Brandt KK, Loit K, Põldmets M, Shanskiy M, Peltoniemi K, Hagner M, Calviño DF. Occurrence, persistence and risk assessment of pesticide residues in European wheat fields: A continental scale approach. Journal of Hazardous Materials. 2025;**494**:138291. DOI: 10.1016/j.jhazmat.2025.138291

[30] EPA. Pesticides Industry Sales and Usage 2008 and 2012 Market Estimates. 2017. https://www.epa.gov/sites/default/files/2017-01/documents/pesticides-industry-sales-usage-2016_0.pdf [Accessed: 2025-September-25]

[31] SpringWell. How widespread and dangerous are pesticides in U.S. drinking water? 2022. https://www.springwellwater.com/dangers-pesticides-in-us-drinking-water/?srsltid=AfmBOor_0gLpEO3VXEb14QVdaRTqQFIipLsL7VZbt9Xvncw5876V3diR [Accessed: 25-September-2025]

[32] Donley N.The USA lags behind other agricultural nations in banning harmful pesticides. Environmental Health. 2019;**18** (1). DOI: 10.1186/s12940-019-0488-0

[33] Singh NK, Sanghvi G, Yadav M, Padhiyar H, Christian J, Singh V. Fate of pesticides in agricultural runoff treatment systems: Occurrence, impacts and technological progress. Environmental Research. 2023;**237**. DOI: 10.1016/j.envres.2023.117100

[34] Beyond Pesticides. A U.S. geological survey finds mixture of pesticides are widespread in U.S. rivers and streams. 2020. https://beyondpesticides.org/dailynewsblog/2020/09/u-s-geological-survey-finds-mixtures-of-pesticides-are-widespread-in-u-s-rivers-and-streams/ [Accessed: 2025-September-29]

[35] Tang FHM, Lenzen M, McBratney A, Maggi F. Risk of pesticide pollution at the global scale. Nature Geoscience. 2021;**14**:206–210. DOI: 10.1038/s41561-021-00712-5

[36] Daily Nation. Toxic dump: The poison killing Lake Victoria. 2020. https://nation.africa/kenya/nation-prime/toxic-dump-the-poisons-killing-lake-victoria-251224#google_vignette [Accessed: 2025-September-29]

[37] Sumon KA, Rashid H, Peeters ETHM, Bosma RH, Van Den Brink PJ. Environmental monitoring and risk assessment of organophosphate pesticides in aquatic ecosystems of north-west Bangladesh. Chemosphere. 2018;**206**:92–100. DOI: 10.1016/j.chemosphere.2018.04.167

[38] Lewis SE, Brodie JE, Bainbridge ZT, Rohde KW, Davis AM, Masters BL, Maughan M, Devlin MJ, Mueller JF, Schaffelke B.Herbicides: A new threat to the Great Barrier Reef. Environmental Pollution. 2009;**157**(8):2470–2484

[39] Wong BBM, Candolin U. Behavioural responses to changing environments. Behavioral Ecology. 2015;**26**(3):665–673. DOI: 10.1093/beheco/aru183

[40] Morrison K, Melhado G, Bose APH, Eastment R, Lagisz M, Manera JL, Michelangeli M, Ozeki S, Wong BBM, Yang Y, Nakagawa S. The impacts of pesticide exposure on fish conspecific interactions: A systematic review and meta-analysis. Environmental Pollution. 2025;**376**. DOI: 10.1016/j.envpol.2025.126353

[41] Wei X, Pan Y, Zhang Z, Cui J, Yin R, Li H, Qin J, Li AJ, Qiu R. Biomonitoring of glyphosate and aminomethylphosphonic acid: Current insights and future perspectives. Journal of Hazardous Materials. 2024;**463**:132814. DOI: 10.1016/j.jhazmat.2023.132814

[42] Benbrook CM. Trends in glyphosate herbicide use in the United States and globally. Environmental Sciences Europe. 2016;**28**:3. DOI: 10.1186/s12302-016-0070-0

[43] Curl CL, Hyland C, Spivak M, Sheppard L, Lanphear B, Antoniou MN, Ospina M, Calafat AM. The effect of pesticide spray season and residential proximity to agriculture on glyphosate exposure among pregnant people in Southern Idaho. Environmental Health Perspectives. 2023;**131**:12. DOI: 10.1289/EHP12768

[44] Hyland C, Spivak M, Sheppard L, Lanphear BP, Antoniou M, Ospina M, Calafat AM, Curl CL. Urinary glyphosate concentrations among pregnant participants in a randomized, crossover trial of organic and conventional diets. Environmental Health Perspectives. 2023;**131**:7. DOI: 10.1289/EHP12155

[45] Lesseur C, Pathak KV, Pirrotte P, Martinez MN, Ferguson KK, Barrett ES, Nguyen RHN, Sathyanarayana S, Mandrioli D, Swan SH, Chen J. Urinary glyphosate concentration in pregnant women in relation to length of gestation. Environmental Research. 2022;**203**. DOI: 10.1016/j.envres.2021.111811

[46] Navarro I, Torre A, Sanz P, Abrantes N, Campos I, Alaoui A, Christ F, Alcon F, Contreras J, Glavan M, Pasković I, Pasković MP, Nørgaard T, Mandrioli D, Sgargi D, Hofman J, Aparicio V, Baldi I, Bureau M, Vested A, Harkes P, Huerta-Lwanga E, Mol A, Geissen V, Silva V, Martínez MA. Assessing pesticide residues occurrence and risks in water systems: A Pan-European and Argentina perspective. Water Research. 2024;**254**:121419

[47] Zago AM, Faria NMX, Fávero JL, Meucci RD, Woskie S, Fassa AG. Pesticide exposure and risk of cardiovascular disease: A systematic review. Global Public Health. 2020;**17** (12):3944–3966. DOI: 10.1080/17441692.2020.1808693

[48] Hladik M, Kolpin DW, Kuivila KM. Widespread occurrence of neonicotinoid insecticides in streams in a high corn and soybean producing region, USA. Environmental Pollution. 2014;**193**:189–196. DOI: 10.1016/j.envpol.2014.06.033

[49] Yan D, Zhang Y, Liu L, Yan H. Pesticide exposure and risk of Alzheimer's disease: A systemic review and meta-analysis. Scientific Reports. 2016;**6**: 32222. DOI: 10.1038/srep32222.PMID: 27581992; PMCID: PMC5007474

[50] Ahmed H, Abushouk AI, Gabr M, Negida A, Abdel-Daim MM. Parkinson's disease and pesticides: A meta-analysis of disease connection and genetic alterations. Biomedicine & Pharmacotherapy. 2017;**90**: 638–649. Epub 2017 Apr 14. PMID:

28412655. DOI: 10.1016/j.biopha.2017.03.100

[51] Lerro CC, Koutros S, Andreotti G, Friesen MC, Alavanja MC, Blair A, Hoppin JA, Sandler DP, Lubin JH, Ma X, Zhang Y, Beane Freeman LE. Organophosphate insecticide use and cancer incidence among spouses of pesticide applicators in the Agricultural Health Study. Occupational and Environmental Medicine. 2015;**72**(10):736–744. DOI: 10.1136/oemed-2014-102798

[52] Liu J, Schelar E. Pesticide exposure and child neurodevelopment: Summary and implications. Workplace Health & Safety. 2012;**60**(5):235–242. DOI: 10.1177/216507991206000507

[53] Sheahan M, Barrett CB, Goldvale C. Human health and pesticide use in Sub-Saharan Africa. The Journal of the International Association of Agricultural Economists. 2017. DOI: 10.1111/agec.12384

[54] Pesticide Action Network (PAN). Parties to the Stockholm Convention agree to phase out the highly toxic pesticide chlorpyrifos. [Press Release]. PAN. 2025. Available from https://panap.net/2025/05/parties-to-the-stockholm-convention-agree-to-phase-out-the-highly-toxic-pesticide-chlorpyrifos [Accessed: 2025-September-29]

[55] The Conversation. Pesticides: Farming chemicals make insects sick at non-deadly doses – Especially in hot weather. 2024. https://theconversation.com/pesticides-farming-chemicals-make-insects-sick-at-non-deadly-doses-especially-in-hot-weather-241856 [Accessed: 2025-September-29]

[56] Vignieri S, Szuromi P, Ash C, Lavine MS, Smith J, Funk MA, Wong W, Matsiko A, Nusinovich Y, Kelly PN, Grocholski B, Isles HM, Charneski C. In Science Journals. Science. 2024;**386**(6720):392–394. DOI: 10.1126/science.adu0446

[57] Gandara L, Jacoby R, Laurent F, Spatuzzi M, Croocker J, et al. Pervasive sublethal effects of agrochemicals on insects at environmentally relevant concentrations. Science. 2024;**386**(6720):446–453. DOI: 10.1126/science.ado0251

[58] WWF. Living Planet Report 2024 – A System in Peril. WWF, Gland, Switzerland: 2024

[59] Ceballos G, Ehrlich PR, Dirzo R. Biological annihilation via the ongoing sixth mass extinction signalled by vertebrate population losses and declines. Proceedings of the National Academy of Sciences of the United States of America. 2017;**114**:E6089–E6096

[60] González-Del-Pliego P, Freckleton RP, Edwards DP, Koo MS, Scheffers BR, Pyron RA, Jetz W. Phylogenetic and trait-based prediction of extinction risk for data-deficient amphibians. Current Biology. 2019;**29**:1557–1563.e3

[61] Wagner DL, Grames EM, Forister ML, Berenbaum MR, Stopak D. Insect decline in the Anthropocene: Death by a thousand cuts. Ecology. 2021;**118**(2):e2023989118. DOI: 10.1073/pnas.2023989118

[62] Food and Agriculture Organization (FAO). Integrated Pest Management: Guidelines for Sustainable Crop Protection. Rome: FAO; 2023

[63] Ali MA, Abdellah IM, Eletmany MR.Towards sustainable management of insect pests: Protecting food security through ecological intensification.

International Journal of Chemical and Biochemical Sciences. 2023;**24**(4): 386–394

[64] Boudh S, Singh JS. Pesticide Contamination: Environmental Problems and Remediation Strategies. In Bharagava R, Chowdhary P, editors. Emerging and Eco-Friendly Approaches for Waste Management. Singapore: Springer; 2019. DOI: 10.1007/978-981-10-8669-4_12 [Accessed: 2025-September-29]

[65] Peshin R, Zhang W, Eddleston M. Integrated pest management in the 21st century: Global perspectives and challenges. Annual Review of Entomology. 2022;**67**:489–511

[66] Choudhury D, Kumar P, Zhimo VY, Sahoo J. Crop rotation patterns and soil health management. In: Kumar P, Srivastav AL, Chaudhary V, van Hullebusch ED, Busquets R, editors. Bioremediation of Emerging Contaminants from Soils. Elsevier; 2024. p. 565–589

[67] Jasrotia P, Kumari P, Malik K, Kashyap PL, Kumar S, Bhardwaj AK, Singh GP. Conservation agriculture-based crop management practices impact diversity and population dynamics of the insect-pests and their natural enemies in agroecosystems. Frontiers in Sustainable Food Systems. 2023;7:1173048. DOI: 10.3389/fsufs.2023.1173048

[68] Karlsson Green K, Stenberg JA, Lankinen Å. Making sense if integrated pest management (IPM) in the light of evolution. Evolutionary Applications. 2020;**13**(8):1791–1805. DOI: 10.1111/eva.13067

[69] Mani M, Natarajan N, Hegde RD, Tej MK. Host plant resistance to insect pests in horticultural crops. Trends in Horticultural Entomology. 2020;335–386. DOI: 10.1007/978-981-19-0343-4_11

[70] Zhou W, Arcot Y, Medina RF, Bernal J, Cisneros-Zevallos L, Akbulut MES. Integrated Pest Management: An update on the sustainability approach to Crop Protection. American Chemical Society. 2024;**9**(40):41130–41147. DOI: 10.1021/acsomega.4c06628

[71] Rejesus RM, Jones MS. Perspective: Enhancing economic evaluations and impacts of integrated pest management farmer field schools (IPM-FFS) in low-income countries. Pest Management Science. 2020;**76**(11):3527–3536. DOI: 10.1002/ps.5912

[72] Oso AA, Oludaisi OC. Effect of Khaya senegalensis and Trichilia emetica plant extracts against sweet potato weevil (Cylas puncticollis Boheman). Asian Journal of Agricultural and Horticultural Research. 2025;**12**(3):210–217. DOI: 10.9734/ajahr/2025/v12i3405

[73] Oso AA, Kolade MB. Evaluating neem extracts, neemsol, and cypermethrin for sustainable insect pest control in cowpea (Vigna unguiculata). Asian Journal of Research in Crop Science. 2025;**10**(3):201–207. DOI: 10.9734/ajrcs/2025/v10i3386

[74] Gundreddy RR, Jayanth BV, Shashikala M, Sibananda D, Gunturi A. Recent Advances in Biopesticide Formulations for Targeted Insect Pest Management: A Review. Journal of Experimental Agriculture International. 2024;**46**(9):876–889. DOI: 10.9734/jeai/2024/v46i92885

[75] Anastasiou E, Fountas S, Voulgaraki M, Psiroukis V, Koutsiaras M, Kriezi O, Lazarou E, Vatsanidou A, Fu L, Bartolo FD, Barreiro-Hurle J,

Gómez-Barbero M. Precision farming technologies for crop protection: A meta-analysis. Smart Agricultural Technology. 2023;(5):100323. DOI: 10.1016/j.atech.2023.100323

[76] The Conversation (2024). Why global pesticide regulation is failing to protect health and the environment. https://theconversation.com [Accessed: 2025-September-29]

[77] European Parliament. Report on the Union's Authorization Procedure for Pesticides (PEST Committee). Brussels: European Parliament Resolution P8 TA; 2019

[78] PAN Europe Activity Report. 2023. Available from https://www.pan-europe.info/resources/reports/2024/06/pan-europe-activity-report-2023 [Accessed: 2025-September-29]

[79] European Commission. EU Bans and Restrictions on Hazardous Pesticides: Progress and Gaps. Brussels: Directorate-General for health and food safety; 2024

[80] Pesticide Action Network (PAN). Hazardous Pesticide Trade and Use in Developing Countries: 2025 Status Report. Hamburg: PAN International Secretariat; 2025

[81] Food and Agriculture Organization (FAO). Global Pesticides Statistics and Trends. Rome: FAO; 2024

[82] World Health Organization (WHO). International Trade Code of Conduct on Pesticide Management: Guidelines for Responsible Use. Geneva: WHO and FAO; 2024

Chapter 3

Pesticide Toxicity to Pollinators: Impacts on Honeybee Growth, Development and Behavior

Lifeng Meng, Jia Xing Huang and Li Jiao Gao

Abstract

Pollinators, particularly bees, are vital to global food production and biodiversity. However, the widespread pesticide use in modern agriculture poses significant threats to their health and survival. This review primarily synthesizes current knowledge on the toxicological impacts of pesticides on bee growth, development, and behavior. We examine key pesticide classes, exposure pathways, physiological and neurological consequences, and species-specific vulnerabilities. Special emphasis is placed on synergistic interactions between pesticides and other stressors such as pathogens and malnutrition. Case studies and policy-related content are also briefly reviewed, including some mitigation strategies such as integrated pest management, biopesticides, and regulatory reforms. Finally, we identify major knowledge gaps and future research directions, advocating for field-realistic testing, ecosystem-level assessments, and cross-disciplinary collaboration to ensure pollinator health and sustainable agriculture.

Keywords: honey bee, pesticide, acute toxicity, chronic toxicity, policy

1. Introduction

Pollinators, especially bees, are fundamental to both the biodiversity of natural ecosystems and agricultural productivity. Approximately 75% of the world's major food crops partially rely on insect pollination, with bees playing a central role [1]. Although bee pollination services are directly tied to global food security, influencing crop yields, quality, and nutritional diversity [2, 3], managed and wild bee populations have been declining across multiple continents in recent years [4].

Numerous factors are implicated in pollinator declines, including climate change, habitat loss, disease and pests, and pesticide application, among which the latter stands out as both pervasive and multifaceted [5]. Pesticides, such as insecticides, herbicides, and fungicides, are widely used in modern agricultural systems to control pests and maximize crop yields. Many of these compounds are applied during the blooming period of crops, directly coinciding with the peak foraging activity of pollinators [6]. Other insecticides exhibit systemic properties; they can be taken up through roots, stems, and foliage to all parts of the plant, including nectar and pollen

[7]. For instance, neonicotinoids, which are designed to target pests with piercing-sucking mouthparts, fall into this category. Even herbicides and fungicides [8], historically considered safe for bees, have been shown to exert indirect or synergistic toxicity under field-relevant conditions [9, 10].

Pesticide toxicity to pollinators typically encompasses both acute and chronic risks. Acute toxicity arises from short-term exposure to high concentrations of pesticides, exemplified by bees foraging on flowers during active spraying events, which may induce immediate mortality or behavioral incapacitation [11]. In contrast, chronic toxicity results from prolonged exposure to low, sublethal concentrations in contaminated nectar, pollen, water, or hive materials, leading to long-term impairments in navigation, learning, reproduction, and immune function [12–14]. These effects are further exacerbated by synergistic interactions under field-realistic conditions, including pesticide mixtures, co-exposure to pathogens (e.g., *Nosema ceranae*), nutritional stress, and environmental temperature fluctuations [4, 15].

Given the pivotal ecological and economic roles of pollinators, assessing pesticide-induced harm to pollinating bees is imperative. While earlier research predominantly focused on immediate lethal effects caused by single compounds, recent findings have highlighted the subtle, cumulative, and synergistic impacts of pesticide exposure on bee development and behavior [16]. These impacts not only threaten individual bees but also imperil colony health, productivity, and long-term survival.

This review synthesizes current knowledge on the types of pesticides relevant to pollinators, the pathways and mechanisms through which bees are exposed, and the documented impacts of pesticides on bee growth, development, and behavior. By integrating findings across toxicology, ecology, and apiculture, this article aims to inform researchers, policymakers, and agricultural stakeholders of emerging pesticide risks and proposes science-backed solutions to protect pollinators.

2. Overview of pesticides relevant to pollinators

2.1 Classification and toxicity

Pollinators are exposed to a diverse range of pesticides, including insecticides, herbicides, and fungicides. Each class exhibits distinct modes of action, environmental persistence, and impacts on pollinator health.

Classification standards for pesticide toxicity to bees are essential for assessing the potential risks of pesticides to bees and hold significant implications for bee protection and the development of new pesticides [17]. Honeybees, particularly *Apis mellifera*, are often used as a surrogate species in pesticide risk assessments [18]. Pesticide toxicity classification primarily relies on acute toxicity data, such as the median lethal dose (LD_{50}) or median lethal concentration (LC_{50}), evaluated across different exposure routes (oral, contact) and life stages (adult bees, larval bees) [19]. Specifically, pesticides with an LD_{50} value of less than or equal to 2 micrograms per bee ($LD_{50} \leq 2$ μg/bee) are considered highly toxic [17]. Representative examples of highly toxic pesticides include neonicotinoids (e.g., imidacloprid, clothianidin, thiamethoxam), which are among the most extensively studied and impactful stressors on honeybee health [20]. Another well-documented highly toxic pesticide is chlorpyrifos, an organophosphate compound that exhibits acute toxicity to honeybees and disrupts their neurological function. In a study on *Apis mellifera* larvae, chlorantraniliprole exhibited high acute oral toxicity to honeybees, with a 48-hour LC_{50} of 6.13 mg active ingredient/L and a 48-hour acute contact LD_{50} of 0.332 μg active ingredient/bee [21]. Pyrethroids have shown variable but significantly detrimental impacts on bee survival in studies; for example, tau-fluvalinate and bifenthrin exhibit efficacy against *Varroa destructor* and relatively lower acute toxicity to bees, and are commonly used as miticides in apiculture [22]. Moderately toxic pesticides generally have an LD_{50} value between 2 micrograms per bee and 10 micrograms per bee (2 μg/bee < LD_{50} ≤ 10 μg/bee) [17]. Examples of such pesticides include acetamiprid, some carbamates, and spirotetramat [18, 23]. Pesticides with low toxicity to bees typically have an LD_{50} value greater than 10 micrograms per bee (10 μg/bee < LD_{50} ≤ 100 μg/bee). This category includes many herbicides, fungicides, and some biological pesticides [24, 25].

Herbicides, often considered less toxic to bees than insecticides, can still have detrimental effects, especially when combined with other stressors, as they can induce high toxicities and systemic physiological disturbances in overwintering honeybees [26]. Glyphosate, the most extensively used herbicide worldwide, has been shown to disrupt the gut microbiota of honeybees, impair nutrient absorption, and immunity [9]. Moreover, herbicide use reduces floral diversity and abundance in agroecosystems, limiting the availability of nutritious forage for both wild and managed bees [26].

Fungicides, though often considered to have low acute toxicity to honeybees, can still pose risks. They are commonly applied during flowering periods under the assumption that they pose minimal risk to bees. However, many fungicides, including chlorothalonil and propiconazole, have been implicated in synergistic toxicity when combined with insecticides, as they inhibit detoxification pathways, such as those involving cytochrome P450 enzymes [10]. Chronic exposure to such mixtures can impair larval development, reduce adult bee longevity, and compromise colony function [26].

2.2 Exposure pathways for bees and pesticide residues

Bees are exposed to pesticides through a complex array of pathways, both inside and outside the hive. These exposures can be direct, indirect, or transgenerational, as detailed below: Outside the hive, foraging bees come into contact with pesticide sprays or residues on treated plant surfaces during active application or shortly thereafter [6]. When bees consume contaminated nectar, pollen, or water, it results in immediate sublethal effects or death, depending on the pesticide's toxicity and dose.

Inside the hive, contaminated resources are subsequently transported back to the hive and incorporated into stored honey, bee bread, and even royal jelly, resulting in colony-wide exposure [27, 28]. Bees, larvae, and the queen come into contact with this contaminated food, and residues in beeswax impair their physiology and behavior over time, causing chronic, cumulative exposure [29]. These contaminants may persist in stored food for extended periods and affect future generations. Additionally, transgenerational impacts have been documented: queen exposure to pesticides can impair egg-laying capacity, and sublethal doses delivered to larvae affect adult behavior and cognitive function [30]. Hence, chronic exposure leads to subtle but significant reductions in colony performance and resilience (**Figure 1**).

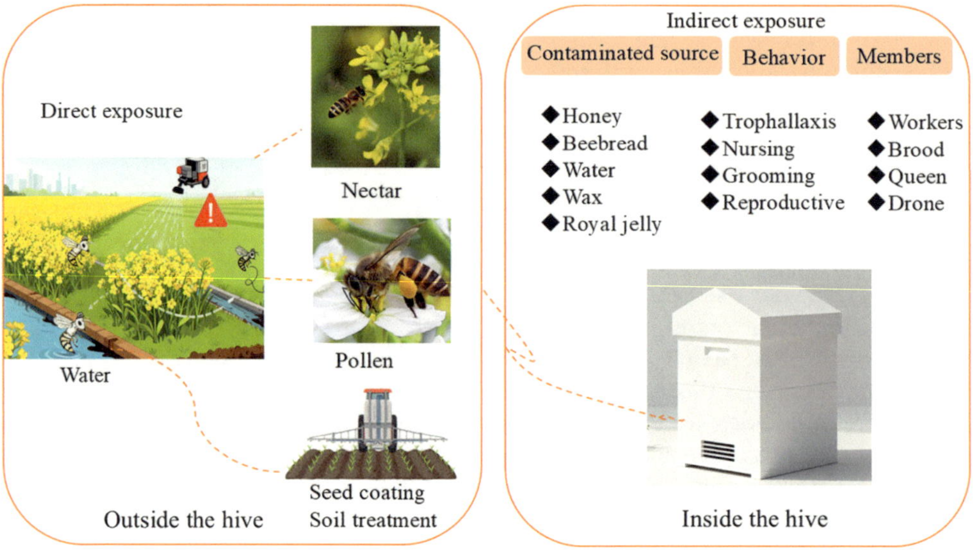

Figure 1.
Exposure pathways of honeybees to pesticides.

Pesticide residues are ubiquitous in various beehive matrices, including pollen, honey, and beeswax [31]. Specific numerical data for pesticide residues vary widely across studies. For example, one study detected τ-fluvalinate concentrations between 14 and 160 ng/g in bees and between 98 and 1,630 ng/g in wax [32]. In another instance, the sum of 10 organochlorine pesticide (OCP) concentrations in honeybee samples was reported as 39.14 ± 11.06 ng/g for urban areas and 39.93 ± 7.09 ng/g for semi-urban areas. Similarly, the sum of 10 OCP concentrations in pollen and honey samples was estimated to be 21.72 ± 4.43 ng/g and 41.83 ± 10.36 ng/g, respectively [33, 34]. Neonicotinoid detection rates in Chinese honey ranged from 35.48% to 83.87%, with mean residue concentrations of 2.39–6.73 μg/kg [35]. Pollen is a crucial route for pesticide entry into the hive and often contains higher concentrations and a greater diversity of pesticide residues compared to honey or beeswax [36]. A study in Western Australia found that contamination was highest in pollen, averaging 2.31 chemicals per sample, followed by beebread (1.69), wax (1.43), and honey (0.32) [36]. Beeswax acts as a long-term reservoir for many lipophilic pesticides due to their persistence and accumulation within the wax matrix [37]. Research in Western Spain analyzed 25 beeswax samples and foundation wax, revealing 18 insecticides, 7 fungicides, and 4 herbicides, including some banned substances [38]. The high levels of miticides and agrochemicals found in North American apiaries often accumulate in hive matrices, including beeswax. Honey, while generally less contaminated than pollen or wax, can still contain pesticide residues [39].

2.3 Amplifying risk factors

In real-world agroecosystems, bees are rarely exposed to a single stressor. Instead, they face a combination of multiple pesticides, pathogens, nutritional deficiencies, and climate extremes. These factors interact in complex and often synergistic ways

[40]. The toxicity of pesticides to bees is influenced not only by dose and chemical composition but also by environmental and biological variables that amplify risk [41]. Environmental modifiers, such as temperature and humidity, can alter pesticide toxicity by affecting metabolic rates or chemical degradation. For example, lower temperatures increase the toxicity of pesticides, posing a risk to overwintering bees [42]. Changes in the ecological environment can also increase bees' sensitivity to pesticides. Monocultures, which promote repeated pesticide application and reduce floral diversity, lead to malnutrition in bees [43, 44].

Synergistic chemical mixtures, where multiple compounds interact to produce effects greater than the sum of their parts, are a significant concern. Co-exposure to different pesticides, such as neonicotinoids and pyrethroids, has been shown to exert synergistic lethality, even when individual concentrations are sublethal [45]. For example, mixtures of imidacloprid, difenoconazole, and glyphosate can induce high toxicity and systemic physiological disturbances in winter honeybees [26]. Binary mixtures of clothianidin with other pesticides also exhibit synergistic toxic effects on honeybees [46]. Additionally, some fungicides inhibit bee detoxification enzymes (e.g., cytochrome P450s), thereby magnifying the toxicity of co-occurring insecticides [47].

Pesticide and pathogen synergies: Numerous studies have demonstrated that pesticide exposure compromises bees' immune systems, increasing their susceptibility to opportunistic infections [48]. For example, neonicotinoids such as imidacloprid and clothianidin have been shown to suppress the expression of immune-related genes, reduce hemocyte counts, and impair melanization and encapsulation responses [49]. One of the most prominent examples is that bees infected with *Nosema ceranae* exhibit increased sensitivity to neonicotinoid-induced mortality, suggesting an immunosuppressive effect and metabolic burden from concurrent stressors [50]. *Nosema ceranae* infection and pesticide exposure collectively contribute to gut dysbiosis, which, in turn, disrupts honeybee physiology [51].

Pesticide exposure can also exacerbate viral infections in bees. Neonicotinoids have been linked to elevated titers of black queen cell virus (BQCV) [52]. Exposure to clothianidin specifically has been shown to negatively impact NF-κB signaling, a crucial pathway for immune responses in honeybees, and this impairment can enhance the proliferation of honeybee parasites and pathogens, such as the deformed wing virus (DWV) [48]. Specifically, clothianidin-induced immunosuppression increases the susceptibility of honeybees to *Varroa destructor* infestation and subsequent DWV replication [53]. These synergistic interactions highlight the necessity of assessing pesticide risks within a broader ecological and immunological framework.

Nutrition and toxicant interactions: pollen diversity and abundance are essential for maintaining bee immunity, detoxification capacity, and colony development [54]. Habitat degradation and intensive monoculture agriculture have significantly reduced floral diversity and foraging options, contributing to chronic nutritional stress among pollinators [54]. Poor nutrition has been shown to exacerbate pesticide toxicity by reducing bees' ability to upregulate detoxification enzymes and antioxidant defenses. Limited access to diverse pollen sources during pesticide exposure further reduces bees' detoxification capacity and resilience. Poor nutrition or pathogen infection (e.g., *Nosema ceranae*), which compromise immune function and reduce detoxification capacity, also play a significant role [55]. For instance, neonicotinoid-exposed bees under poor nutritional conditions had lower survival and impaired recovery compared to nutritionally replete counterparts [11], indicating that pesticide toxicity cannot be considered in isolation but must be evaluated alongside ecological context.

Summary: These findings emphasize that pesticide risk assessment frameworks must account for chronic, low-dose, and combinatorial exposures, as well as interactions with biological and environmental co-stressors. Standard acute toxicity metrics (e.g., LD_{50}) fail to capture these real-world dynamics.

3. Impacts on bee growth and development

Pesticide exposure, both acute and chronic, has been demonstrated to adversely affect multiple stages of honeybee development. These effects extend from the larval and pupal stages to adult glandular development, neurodevelopment, metabolic processes, and reproductive capacity, ultimately jeopardizing colony health and pollination services.

3.1 Larval and pupal development

Exposure to pesticides during early developmental stages has profound implications for the viability of larvae and pupae. Both laboratory and field studies consistently report increased mortality, developmental delays, and morphological deformities in exposed broods.

Neonicotinoids and pyrethroids have been shown to cause high mortality and abnormal development in bee larvae [16]. Environmentally relevant concentrations (0.7–377 ppb) of imidacloprid disrupt larval molting and perturb endocrine homeostasis, leading to incomplete or arrested development [56]. *Apis mellifera* larvae exposed to a field concentration of thiamethoxam (0.001 ng/µL) exhibited effects on larval and pupal survival and adult emergence rates, alongside cytotoxic effects on the digestive cells of the midgut [57]. Organophosphorus pesticides, like chlorpyrifos, also pose a substantial threat. Larvae of the stingless bee Scaptotrigona bipunctata exposed to chlorpyrifos developed into lighter, smaller, and deformed adult workers [58].

Glyphosate (GLY) is one of the most widely applied herbicides and has been reported to delay honeybee eclosion. *In vivo* experiments reveal that GLY prolongs the larval stage of honeybees by modulating the expression of gene immune and detoxifying genes [59]. Transcriptome analysis has confirmed that the expression of genes associated with defense and metabolism changes following exposure to GLY [60]. Many pesticides, such as chlorpyrifos, imidacloprid, amitraz, fluvalinate, coumaphos, myclobutanil, chlorothalonil, glyphosate, and simazine, induce cell death in the midgut, salivary glands, and ovaries of honeybee larvae.

During pupation, sublethal pesticide exposure can delay eclosion, reduce adult emergence rates, and induce structural abnormalities, including deformed wings and underdeveloped thoracic segments [61]. These defects are often linked to DNA damage, impaired protein synthesis, and endocrine disruption.

3.2 Adult growth inhibition and metabolic disruption

Pesticides also impair adult honeybee development by disrupting nutrient assimilation and energy metabolism. As a result, exposed bees often exhibit reduced body weight, smaller body size, and shortened lifespan. The mechanism by which pesticides impact the development, growth, and behavior of adult honeybees is consistent with those affecting larvae, primarily through disrupting metabolism, hormonal systems, and immune functions [62].

Consequently, low body weight and insufficient energy reserves accelerate honeybee development, causing precocious behavioral maturation, where nurse bees transition prematurely to foragers, shortening their lifespan, and rendering individuals more vulnerable to environmental stressors, ultimately compromising colony stability.

3.3 Gland and organ development

Pesticides severely impact the glands of honeybees at multiple physiological levels, with larval-stage exposure exerting the most profound and irreversible effects on the development of specific glands [63]. Currently, research on the effects of pesticides on bee glands and organs focuses primarily on the hypopharyngeal gland (HPG, responsible for royal jelly secretion), fat body (involved in nutrient storage and detoxification), midgut (responsible for food digestion), and brain [64]. Significant advancements have been made in understanding these effects through histopathological analysis, ultrastructural studies, biochemical profiling, and investigation into molecular mechanisms.

Neonicotinoid insecticides (e.g., thiamethoxam) and pyrethroids (e.g., lambda-cyhalothrin) exert significant adverse effects on the morphology and function of honeybees [65]. It has been reported that thiamethoxam significantly reduces acinus diameter and impairs gland maturation [66]. Pyrethroids (e.g., lambda-cyhalothrin) dysregulate calcium channel-mediated neurotransmission, inhibit acetylcholine signaling, and disrupt HPG secretory cycles, thereby impairing royal jelly synthesis and accelerating the nurse-to-forager transition [64, 67]. Conversely, cyflumetofen induces acinus dilation with secretion accumulation in the distal apparatus of HPG acini, triggering mTOR-mediated feedback suppression of protein synthesis [68]. This pathological retention further reduces glandular output and compromises brood care capacity. Lambda-cyhalothrin also triggers dilation of the rough endoplasmic reticulum in secretory cells and impairs gland maturation. Furthermore, thiamethoxam activates oxidative stress cascades in the brain, upregulating heat shock proteins (HSP70/HSP90) and disrupting the glutamate/GABA neurotransmitter balance, which impairs neuronal regulation of gland maturation and indirectly suppresses HPG secretion ability [69]. Thiamethoxam also downregulates tubulin α-1 expression in mushroom bodies, destabilizing microtubule networks, impairing vesicular transport in HPG acini, and reducing secretory granule exocytosis and royal jelly production [69].

Regarding the midgut, teflubenzuron activates caspase-dependent apoptosis in midgut cells, resulting in cytoplasmic vacuolization, sloughing of cellular fragments, increased autophagosome formation, and microvilli disruption in digestive cells [70]. Glutathione S-transferase (GST) participates in phase II detoxification; exposure to high-concentration teflubenzuron ($\geq LC_{50}$) suppresses GST activity, leading to the inhibition of chitin biosynthesis and disorganization of the peritrophic matrix in the midgut, which exacerbates midgut vulnerability to pathogen invasion [71].

3.4 Neurodevelopmental disruption

Exposure to sublethal concentrations of pesticides severely impairs the development of the honeybee nervous system, especially neonicotinoids. These compounds interfere with synaptogenesis and axonal growth, particularly in brain regions critical for cognitive function [72]. Synapsis is critical for regulating neurotransmitter release, synaptic signal transmission, synaptogenesis, plasticity, and remodeling. Larval exposure to sublethal concentrations of thiamethoxam significantly reduces synapsin levels in the mushroom bodies (MBs) and antennal lobes (ALs) of honeybees during both the pupal stage and in newly emerged worker bees, indicating that diminished synapsin in MBs and ALs may induce neurophysiological disruptions in honeybees, thereby compromising neural functions such as olfactory learning and memory consolidation [73]. Exposure to imidacloprid and clothianidin during larval stages has been shown to reduce the volume of mushroom bodies, the centers for learning and memory in the insect brain [74]. Additionally, pesticide exposure alters the expression of genes involved in neurotransmission, including those in acetylcholine and dopamine signaling pathways [75]. Sublethal doses of clothianidin alter nAChRα7 expression, reducing responsiveness to sucrose solutions [76], impairing the consolidation and retrieval of honeybee memory [77], and contributing to long-term deficits in navigation, communication, and foraging efficiency [78].

3.5 Growth inhibition and metabolic disruption

Pesticides significantly impair honeybee growth and metabolic functions, leading to a range of detrimental effects [4]. Sublethal pesticide doses disrupt metabolic processes in honeybees, resulting in impaired growth and development. Chronic exposure to neonicotinoids affects metabolic physiology, impacting nutritional and energetic homeostasis. For instance, exposure to low concentrations of clothianidin lowers protein, lipid, carbohydrate, and glycogen levels in honeybees [62]. Imidacloprid suppresses the expression of the molting hormone gene 20-hydroxyecdysone (20E), downregulates *Br–c* and *JHAMT* genes (thereby disrupting larval molting), and upregulates the *vg* gene to perturb endocrine homeostasis, leading to incomplete or arrested development [56]. Additionally, imidacloprid disturbs the expression of development-related genes in honeybees, such as those involved in the insulin-signaling pathway and Hippo-signaling pathway, both of which are critical for growth regulation. Similarly, chlorpyrifos, cypermethrin, and thiacloprid can alter the expression of metabolism-related genes, such as those encoding vitellogenin, major royal jelly proteins 2 and 3, insulin-like peptide (ilp1), and alpha-glucosidase [79]. Pesticide exposure alters the expression of heat shock proteins (HSPs), which are critical for cellular protein homeostasis [80], and these alterations compromise the bees' ability to digest pollen and metabolize nutrients [81]. Pesticides also suppress key metabolic enzymes (including cytochrome P450, GST, and esterases) involved in detoxification and lipid metabolism [82], disrupt the carbohydrate metabolic pathway (crucial for energy production), and impair essential functions in bees [83]. For example, pesticides can inhibit carbonic anhydrase, an enzyme vital for various physiological functions, including pH regulation and ion transport [84]. Furthermore, pesticides alter the gut microbiome community, which is important for nutrient absorption and immune development

[85]. Disruptions to the gut microbiota can lead to poor nutrient absorption and increased susceptibility to diseases [9], ultimately causing reduced body weight, smaller body size, and shortened lifespan.

3.6 Reproductive toxicity in queens and drones

Pesticide exposure, even at sublethal, field-relevant doses, impairs the reproductive health of these crucial castes, threatening colony survival and productivity [86]. The reproductive castes of honeybee colonies – queens and drones – are highly sensitive to pesticide exposure [86]. In queens, direct exposure during development to pesticides like neonicotinoids or fungicides like boscalid can impair reproductive anatomy, such as increasing ovariole number, reducing the quality and quantity of stored sperm, diminishing ovary development, reducing ovary size, declining the quality of the eggs and egg eclosion [20, 87], increasing mortality during nuptial flights, and decreasing egg-laying rates [88]. Neonicotinoids can also affect a queen's physiology and survival by depressing the expression of genes associated with detoxification and energy metabolism, reducing their immune capacity, and increasing susceptibility to diseases [89]. Furthermore, queen honeybee pheromones and reproductive behavior can be affected by pesticide exposure during development [90]. Exposure to miticides such as amitraz has been shown to paradoxically increase mating frequency, possibly as a compensatory behavior for physiological compromise [91]. These impairments directly affect colony renewal and longevity.

In drones, neonicotinoid exposure is associated with reduced sperm count, motility, and viability [92]. Miticides also adversely affect the survival rate and sperm viability of drones [93]. Additionally, behavioral deficits during mating flights, such as impaired orientation and lowered competitiveness, have been observed [94, 95]. These effects result in poor queen fertilization and an increased risk of colony failure.

3.7 Colony-level and ecological consequences

Developmental and reproductive impairments in individual bees ultimately scale to population-level consequences. Decreased worker emergence, reduced foraging efficiency, and impaired brood care contribute to colony depopulation and lowered productivity [78].

The cumulative stress from pesticide exposure has been implicated in colony collapse disorder (CCD), a phenomenon marked by the sudden disappearance of worker bees and subsequent hive collapse [96]. Furthermore, declines in colony strength compromise pollination services, threatening plant reproduction, biodiversity, and agricultural yields [2, 4] (**Figure 2**).

4. Behavioral disruptions induced by pesticides

Pesticide exposure, particularly at sublethal levels, has been shown to impair essential behaviors in honeybees (*Apis mellifera*) and other pollinators. These disruptions span across individual navigation and foraging, brood care, colony-level communication, and innate behaviors involved in social immunity [97]. Given that

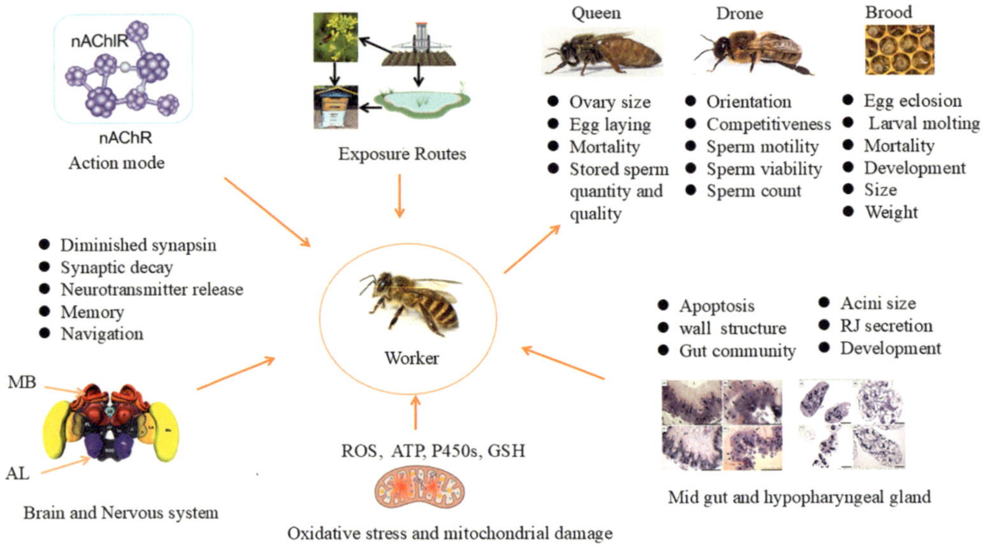

Figure 2.
Mode of action and sublethal effects of neonicotinoid pesticides on honeybees (Adapted from Ahsan, 2025, biology, "The sublethal effects of Neonicotinoids on honeybees").

pollinator behavior is tightly integrated with ecosystem functioning and colony survival, even subtle behavioral impairments can lead to cascading ecological and economic consequences.

4.1 Foraging behavior and navigation

Foraging efficiency and spatial orientation are among the most consistently disrupted behaviors in bees following pesticide exposure [98]. Bees rely heavily on visual landmarks, the sun compass, and memory of spatial cues to navigate between the hive and floral resources [99]. These functions are particularly sensitive to neurotoxic pesticides, such as neonicotinoids and organophosphates [78].

Spatial memory and orientation disruption have been demonstrated in bees exposed to imidacloprid and clothianidin, which interfere with cholinergic signaling in the central nervous system [12]. Foragers frequently fail to return to the colony, resulting in increased forager loss and reduced pollen and nectar influx [100]. Foraging activity and efficiency also decline significantly upon pesticide exposure. Bees exposed to thiamethoxam or sulfoxaflor exhibit delayed initiation of foraging trips, shortened foraging durations, and reduced flower visitation rates [101, 102]. Their ability to learn and remember the location of food sources, as well as to follow pheromone trails, is interfered with. These changes are not merely due to physical weakness but result from pesticide-induced impairments in motivation, sensory integration, and neuromuscular coordination [12].

Altered floral preferences have been observed in bees after exposure to certain pesticides. Land-use intensity and management practices can change the composition of potential host plants, indirectly affecting pollinator behavior by reducing flower cover and diversity [103]. Surprisingly, bees may even develop a preference for pesticide-laced nectar, a phenomenon hypothesized to involve reward pathway hijacking [104]. This maladaptive preference further increases their toxic burden and disrupts normal foraging ecology.

4.2 Social and communication behaviors

The social organization of honeybees depends on precise communication and cooperation. Pesticides interfere with these fundamental processes, weakening the colony's ability to allocate labor, coordinate foraging, and maintain homeostasis.

Waggle dance impairment, a critical mechanism for conveying spatial information about food sources, is among the most studied effects. Bees exposed to neonicotinoids show shorter dance durations, lower frequencies of dances, and less precise vector information [105]. This degradation in communication leads to decreased recruitment efficiency and reduced colony foraging success.

Trophallaxis, the mouth-to-mouth transfer of food and chemical cues, is crucial for colony health and communication [106, 107]. However, pesticide exposure can disrupt this vital behavior by affecting antennal sensitivity and motor control, thereby reducing the frequency and accuracy of trophallactic events [108]. A study on native Indian honeybees (*Apis cerana*) found that pesticide exposure impaired their olfaction, leading to a significant decrease in the proboscis extension response, which is essential for feeding and trophallaxis [109]. This impairment affects the distribution of food, pheromones, and immune-regulating compounds within the colony.

4.3 Immune-related behaviors

Honeybees exhibit a suite of behavioral adaptations collectively termed "social immunity," including hygienic behaviors, self-grooming, allogrooming, and social distancing, that help contain infections and parasite loads within colonies. Pesticide exposure compromises these behaviors, indirectly facilitating disease transmission.

Hygienic behaviors, including brood grooming and the removal of diseased larvae, are essential for colony-level immunity [110]. This behavior helps to prevent the spread of diseases and parasites within the colony, contributing to overall colony health and survival [111]. Exposure to sublethal doses of pesticides negatively impacts hygienic behavior in honeybees; for instance, coumaphos and fluvalinate reduce hygienic responsiveness and delay the detection of infected brood [112]. It can also affect the neurological processes that mediate hygienic behavior [113], leading to an increase in the prevalence of brood diseases within the colony. However, these pesticides are commonly used to control *Varroa* mites in honeybee colonies [114], increasing colony vulnerability to pathogens and parasites.

Grooming behavior is another important immune-related behavior of honeybees and is significantly suppressed in bees exposed to neurotoxicants. Affected individuals display decreased responsiveness to irritants and spend less time grooming themselves or nestmates [115], while some studies suggest that pesticides have no effect on the grooming behavior of honeybees; they may even enhance the bees' sensitivity to foreign substances and increase their grooming frequency [116, 117].

Together, these behavioral deficits compromise both the individual and collective immune defenses of the colony, rendering them more susceptible to multiple stressors (e.g., parasitic mites like *Acarapis woodi*) and accelerating population decline.

5. Integrated case studies and mechanistic insights

While individual studies have delineated the toxic effects of pesticides on various aspects of bee physiology, behavior, and colony dynamics, a comprehensive understanding of their synergistic and system-wide impacts is critical. This section synthesizes key case studies with mechanistic insights to demonstrate how specific pesticide exposures translate into population declines and ecological instability through multi-layered biological pathways.

5.1 Representative case study: Neonicotinoid-linked mass mortality in Europe

One of the most well-documented and impactful events illustrating the consequences of pesticide exposure on pollinators is the mass mortality of honeybee colonies in several European countries during the late 2000s. In 2008, nearly 12,000 honeybee colonies in Baden-Württemberg, Germany, suffered acute collapse following maize sowing treated with neonicotinoid-coated seeds, primarily clothianidin [96, 118]. Mechanical abrasion during sowing released pesticide-laden dust into the environment, contaminating surrounding flora and exposing foraging bees to lethal doses.

Subsequent investigations revealed high concentrations of clothianidin in pollen loads and bee tissues collected from affected colonies. Behavioral impairments, such as disorientation, reduced homing success, and muscle tremors, were observed, suggesting neurotoxicity consistent with neonicotinoid exposure [119]. This incident significantly influenced European Union (EU) regulatory policy, leading to a partial ban of three major neonicotinoids in 2013 and a full ban on outdoor agricultural uses by 2018 [120].

This case underscores the complexity of pesticide-related risks: exposure does not solely depend on application dosage, but also on delivery methods (e.g., seed coatings), environmental dissemination (e.g., dust drift), and the foraging ecology of bees. It also highlights the importance of assessing indirect exposure routes in risk assessments.

5.2 Toxicological mechanisms underpinning bee declines

5.2.1 Molecular targets and oxidative stress

Pesticides can negatively impact honeybees in various ways, affecting their behavior, physiology, and overall colony health. Neonicotinoids, such as imidacloprid and clothianidin, act as agonists of insect nicotinic acetylcholine receptors (nAChRs), competitively inhibiting acetylcholine binding and perturbing cholinergic neurotransmission in the central nervous system, thereby disrupting signal transduction and impairing crucial behaviors like navigation, foraging, mating, and communication [121]. Imidacloprid, at field-realistic doses, partially induces neuronal inactivation in higher brain regions such as the mushroom body (MB), impairing honeybee learning and memory, while high concentrations (e.g., 1 mM imidacloprid) antagonize γ-aminobutyric acid (GABA) receptors, exacerbating neural excitability imbalances [122]. Secondly, neonicotinoid pesticides impair bee peripheral perception, as thiacloprid at 4,500 ppb alters antennal sensitivity to specific odorant molecules (e.g., benzyl alcohol, 1-hexanol), though analogous effects for imidacloprid remain debated. Imidacloprid interacts with odorant-binding proteins (OBPs), hindering odorant molecule delivery to olfactory receptors. Regarding central olfactory coding, calcium imaging reveals that imidacloprid perfusion (2.5 ppm) reduces response intensity in antennal lobe (AL) projection neurons, diminishing odor discriminability and mirroring deficits in behavioral assays such as the proboscis extension reflex (PER) learning paradigm. Furthermore, imidacloprid at 0.8 ng/bee induces optic lobe cell death, potentially compromising visual navigation capabilities [72, 123]. Thirdly, pesticide-induced oxidative stress causes reactive oxygen species (ROS) to accumulate beyond the detoxification capacity of bees' antioxidant systems (e.g., SOD, CAT), leading to mitochondrial dysfunction, DNA fragmentation, and apoptosis [124, 125].

5.2.2 Endocrine disruption and developmental inhibition

Several insect growth regulators and hormone analogs used as pesticides interfere with juvenile hormone (JH) and ecdysone signaling pathways. Compounds such as methoprene and pyriproxyfen act as JH mimics, disrupting the larval-pupal transition and leading to developmental arrest or morphological abnormalities in pupae [126].

Chronic exposure to such endocrine-disrupting compounds reduces reproductive potential by impairing queen oviposition and reducing drone sperm viability [92, 127]. Exposure to insect growth disruptors (IGDs) affects honeybee reproductive behaviors and ovarian protein expression [128]. Developmental exposure to hormone-mimicking IGDs can alter the expression of endocrine-related genes in worker honeybee brains and hypopharyngeal glands [129]. These glands are responsible for producing royal jelly, which is crucial for larval development and queen bee rearing, leading to decreased colony renewal capacity and eventual population collapse. These physiological perturbations reflect broader dysregulation of neuroendocrine axes, ultimately translating into colony-scale instability.

6. Species-specific vulnerability

Not all bee species are equally affected by pesticide exposure. Differences in life history traits, social organization, detoxification capacity, and foraging behavior contribute to species-specific vulnerability, raising concerns about generalizing toxicity data from a few model organisms to the entire pollinator community.

6.1 Honeybees vs. bumblebees vs. solitary bees

*6.1.1 Honeybees (*Apis mellifera*)*

Apis mellifera is the most studied pollinator species and serves as the standard model in regulatory toxicity testing. While honeybees do suffer from pesticide exposure, their eusocial structure and large colony size provide a degree of buffering against individual losses. Division of labor and redundancy in worker roles can compensate for impaired individuals to some extent [78].

Additionally, honeybees possess relatively effective detoxification mechanisms, including upregulated cytochrome P450 enzymes, allowing them to metabolize certain insecticides more efficiently than other species [130, 131].

*6.1.2 Bumblebees (*Bombus *spp.)*

Bumblebees exhibit greater sensitivity to many pesticides, partly due to their smaller, annual colony character and limited redundancy in labor. Their smaller colony size (hundreds rather than thousands of individuals) means they have less capacity for social buffering compared to honeybees. Sublethal impairments in a few individuals can have outsized impacts on colony foraging success and brood development. Studies have shown that neonicotinoids can cause colony-level sublethal effects on bumblebees at lower concentrations than those for honeybees [132].

Exposure to environmental contaminants, such as particulate matter, also affects both species, though *Apis mellifera* has historically been the primary focus as a pollution indicator. However, recent research suggests that bumblebees like *Bombus terrestris* are also significant accumulators of particulate matter and associated trace elements. It has also been shown to have higher accumulation of pesticide residues in brain tissue and lower expression of detoxification genes compared to honeybees [133]. Consequently, even sublethal effects on individual bumblebees can have a more pronounced impact on overall colony growth, reproduction, and survival [134].

Pesticide ingestion can widely alter the social behavior of bumblebees within the nest, impairing nursing, colony thermoregulation, and their social dynamics. Bumblebees' foraging range varies by species, but they generally forage closer to the nest than honeybees. However, this does not necessarily reduce their exposure risk; they can still encounter pesticides in agricultural areas, and their specific floral preferences might expose them to different types and concentrations of chemicals [135]. Chronic sublethal pesticide exposure can affect their ability to collect resources efficiently and their homing ability [136]. Furthermore, such exposure can alter the interactions between bumblebees and wild plants, potentially impacting pollination services.

Fewer studies have focused on bumblebee's detoxification capacity. It has been suggested that bumblebees may generally exhibit a lower detoxification capacity than honeybees [137], potentially making them more susceptible to certain stressors, such as neonicotinoids. A preliminary evaluation comparing the sensitivity of *Bombus terrestris* and *Apis mellifera* to insecticides, fungicides, and herbicides indicated lower or similar contact and oral median lethal dose (LD50) values for bumblebees compared to honeybees, suggesting higher or comparable sensitivity in bumblebees depending on the chemical [138, 139].

6.1.3 Solitary bees

Solitary bees, the largest group of bees worldwide, play important roles in the biodiversity of ecosystems and agriculture [140]. However, they are uniquely vulnerable to pesticide exposure due to their distinct life histories, nesting strategies, and foraging behaviors [141]. Many solitary bees are ground-nesting or cavity-nesting, creating nests in soil or pre-existing cavities. This means they can be exposed to pesticides present in the soil, on plant surfaces, or within the materials they use to construct their nests [143]. One of the primary routes of exposure for solitary bees is through contaminated pollen and nectar collected from agricultural fields [143]. Solitary bee larvae are particularly susceptible as they develop on provisions (pollen and nectar mixed with secretions) stored in their nests [144]. Studies have demonstrated that ingestion of provisions contaminated with pesticides can increase larval mortality and decrease the successful emergence of adult solitary bees in the subsequent year [143]. Nesting materials, such as leaves and soil, can also transfer pesticides into food provisions, leading to additional oral and contact exposure for larvae [142].

Neonicotinoids are widely recognized for their detrimental effects on bees. Even at field-realistic concentrations, neonicotinoid exposure has been shown to affect foraging efficiency, nesting behavior, and reproductive success in ground-nesting solitary bees [123]. The neonicotinoid acetamiprid and the fungicide azoxystrobin have been shown to affect the locomotion activity and mushroom bodies of the Neotropical solitary bee *Centris analis*, indicating neurotoxic effects [145]. Even novel insecticides like sulfoxaflor have been shown to have negative impacts on the survival rate and foraging performances of solitary bees [146]. Co-exposure to multiple pesticides or other stressors, such as insecticides and fungicides, can lead to synergistic or additive negative effects that are more severe than individual exposures [147]. Fungicide-insecticide co-exposure can affect nest recognition and parental investment in solitary bees, highlighting impacts on complex behaviors critical for reproduction [148]. Nutritional stress and climate change can also exacerbate the negative impacts of pesticides on solitary bees [149].

Solitary bees are more sensitive to pesticides than honeybees. Flupyradifurone (FPF), a butenolide insecticide with a similar mode of action to neonicotinoids, has been deemed "practically non-toxic to adult honeybees" in risk assessments, yet it can alter behavior and offspring production in solitary bees like *Xenoglossa pruinosa* [150]. *Megachile rotundata* is more than 2,500 times more sensitive to thiacloprid and 170 times more sensitive to FPF than honeybees [151]. This may be due to their unique life cycle, nesting strategies, and fewer detoxifying enzyme genes, such as those of the P450 family and carboxylesterases, which are significantly contracted in *Megachile rotundata* [152, 153].

6.2 Model organisms and research bias

Despite the diversity of pollinator species (>20,000 globally), pesticide toxicology research remains heavily skewed toward *Apis mellifera*, which constitutes over 80% of published bee toxicology studies [154]. This overreliance limits ecological relevance, especially since wild bees often have distinct foraging behaviors, phenologies, and physiological traits [155]. Emerging studies emphasize the urgent need to incorporate non-Apis pollinators into pesticide risk assessments. Regulatory frameworks, such as those proposed by the EU Food Safety Authority (EFSA), are beginning to acknowledge this gap and recommend the inclusion of representative species from different ecological guilds [156]. Greater inclusion of diverse bee taxa in toxicological research will enable more accurate risk predictions and guide conservation strategies tailored to the vulnerabilities of specific pollinator groups. The ultimate goal is to move toward a comprehensive, multi-species risk assessment framework that accurately reflects the diverse vulnerabilities of the entire pollinator community, ensuring adequate protection for these essential ecological and agricultural service providers.

7. Mitigation strategies and policy implications

The escalating risks posed by pesticides to pollinators necessitate an integrated mitigation framework that combines scientific innovation, sustainable agricultural practices, and progressive policy interventions. This section explores current strategies and future directions for reducing pesticide impacts on pollinators through improved risk assessment, adoption of sustainable alternatives, implementation of integrated pest management (IPM), and regulatory reform.

7.1 Risk assessment improvements

7.1.1 Incorporation of sublethal and behavioral endpoints

Traditional pesticide risk assessments have focused predominantly on acute toxicity measures, such as the lethal dose 50 (LD_{50}) in adult *Apis mellifera*. While valuable, such metrics often fail to capture the full spectrum of sublethal and behavioral effects that are critical for colony performance and survival. Sublethal impairments, such as deficits in navigation, foraging, brood care, and reproduction, have been increasingly documented even at environmentally relevant concentrations (Desneux et al., 2007; Cresswell, 2011). Regulatory bodies, such as the European Food Safety Authority (EFSA, 2013) and the U.S. Environmental Protection Agency (USEPA, 2014) are gradually incorporating endpoints such as Proboscis Extension Reflex (PER) tests, queen production rates, and colony thermoregulation to better reflect real-world risks.

7.1.2 Simulation of field-realistic exposure

Standard laboratory assays often fail to simulate the complexity of pesticide exposure under natural field conditions, where bees encounter fluctuating concentrations, mixed compounds, and environmental stressors such as nutritional deprivation or parasitism. Semi-field (e.g., tunnel tests) and full-field trials that simulate chronic, low-dose exposure in combination with other biotic and abiotic stressors are increasingly recognized as more ecologically valid (Rundlöf et al., 2015). Exposure models such as BeeREX integrate multiple exposure pathways (oral, contact) with pesticide residue data and colony behavioral parameters to predict long-term risks under realistic conditions (USEPA, 2014).

7.1.3 Deployment of monitoring and management tools

A critical component of this integrated approach is the refinement of environmental risk assessment (ERA) methodologies for pesticides. Current ERA frameworks exhibit shortcomings, especially in addressing sublethal and microbiota-mediated effects [157]. For example, incorporating bee gut microbiota as a biomarker-based monitoring tool can significantly enhance the accuracy of environmental risk assessments [158]. New approach methods (NAMs) are expected to be a main component of future integrated approaches to testing and assessment, integrating mechanistic insights, exposure science, and advanced analytical techniques [159]. These advancements will allow for a more comprehensive understanding of pesticide impacts, moving beyond acute toxicity to include long-term, chronic, and synergistic effects on non-target organisms [160]. A global nectar and pollen pesticide residue database is also being developed to provide essential data for pollinating bee dietary risk assessment, accounting for variables like crop attractiveness, pesticide physicochemical properties, application methods, and timing [161]. Technological advances are also enabling the deployment of real-time pesticide monitoring systems. Portable biosensors and immunoassay strips allow rapid detection of pesticide residues in pollen, nectar, and hive matrices [162]. When integrated with GIS-based mapping and decision-support systems (DSS), these tools can help optimize pesticide application schedules to avoid periods of peak foraging, thereby reducing bee exposure.

7.2 Sustainable alternatives to conventional pesticides

7.2.1 Adoption of biopesticides

The shift toward sustainable agriculture heavily relies on the adoption of eco-friendly alternatives and practices that reduce reliance on synthetic pesticides. Biopesticides, derived from natural materials such as plants, bacteria, or fungi – offer a promising alternative to conventional synthetic pesticides [163]. Botanical insecticides (e.g., pyrethrins from *Chrysanthemum cinerariifolium*) and microbial agents (e.g., *Bacillus thuringiensis, Beauveria bassiana*) are characterized by rapid environmental degradation, target specificity, and low toxicity to pollinators [164, 165]. However, challenges such as low environmental persistence, reduced shelf life, and formulation instability remain and require improved delivery and stabilization techniques.

7.2.2 Dose optimization and smart delivery technologies

The principle of "low dose, high efficiency" is being implemented via novel formulation and application strategies. Mesoporous nanomaterials present transformative solutions by enhancing pesticide stability, bioavailability, and controlled release, addressing issues such as environmental residues and pest resistance associated with conventional pesticides [139]. Precision agriculture technologies, including drone-assisted spraying, electrostatic droplet deposition, and variable-rate application systems, further reduce drift into pollinator habitats and improve pesticide application efficiency.

7.3 Integrated pest management and agroecological practices

7.3.1 Core IPM principles

Integrated pest management is a cornerstone of sustainable agriculture, aiming to control pests while minimizing reliance on chemical pesticides and promoting ecosystem health, representing a holistic, knowledge-intensive approach that integrates biological control, cultural practices (e.g., crop rotation), habitat manipulation, and threshold-based pesticide use [166, 167]. Empirical studies consistently demonstrate that IPM strategies can significantly decrease pesticide usage without negatively impacting crop yields, fostering sustainable intensification of agriculture [168]. The EU's approach involves IPM strategies, which aim to reduce reliance on chemical pesticides, and has set ambitious targets, such as a 50% reduction in pesticide use by 2030 as part of its European Green Deal. Moreover, IPM practices significantly lower risks to non-target organisms, including bees, by reducing the frequency, intensity, and spectrum of pesticide exposure [169]. The concept of Integrated Pest and Pollinator Management (IPPM) expands IPM; this framework emphasizes effective pesticide risk assessment within a broader strategy that considers both pest control and pollinator protection [170]. This includes developing methods to optimize pesticide application timing to minimize exposure during periods of high pollinator activity [171].

7.3.2 Landscape-level practices to support pollinators

Agroecological landscape design can synergize with IPM principles to bolster pollinator health. Flower strips, hedgerows, and intercropping not only provide nutritional resources for pollinators but also support populations of natural enemies that suppress pest populations [172]. Temporal staggering of pesticide applications to avoid peak foraging hours (e.g., spraying during early morning or late evening) has also proven effective in minimizing direct exposure risks.

7.4 Legislative and policy reforms

7.4.1 Regulatory precedents and frameworks

The EU's ban on the outdoor use of neonicotinoids (imidacloprid, clothianidin, and thiamethoxam) in 2018 marked a turning point in pesticide regulation aimed at pollinator protection [173]. Other countries, including Canada and France, have followed suit with restrictions, enhanced pollinator risk assessments, and mandatory buffer zones around flowering crops. Regulatory innovations increasingly include requirements for pollinator labeling, exposure simulation, and pre-application notification systems [174, 175]. In the US, a federal pollinator protection plan was established in 2015. Many states have subsequently developed their own plans, but many state plans lack policy elements that address monitoring, evaluation, and adjustment [176].

7.4.2 Incentivizing pollinator-friendly farming

Agri-environmental subsidies and market-based certification schemes play an essential role in shifting farmer behavior to adopt pollinator-friendly practices, aiming to mitigate pollinator declines and enhance agricultural sustainability [177, 178]. Programs like USDA Organic, BeeBetter Certified™, and the EU's Common Agricultural Policy (CAP) serve as prominent examples of agri-environmental subsidy programs. CAP measures, particularly those under Ecological Focus Areas and agri-environmental and climate measures, aim to support biodiversity, including wild pollinators, on farmland [179, 180].

7.4.3 International collaboration and knowledge transfer

Given the migratory behavior of certain pollinators and the global trade of agricultural commodities, international cooperation is indispensable. Platforms such as the Intergovernmental Science-Policy Platform on Biodiversity and Ecosystem Services (IPBES) and the FAO's Global Pollination Project promote cross-border harmonization of pesticide regulations and facilitate technical knowledge exchange [181]. Aligning data standards, risk assessment protocols, and enforcement mechanisms will be vital to managing transboundary threats to pollinators.

7.5 Limitations

While regulatory frameworks, such as the EU's ban on neonicotinoids, reflect incremental progress in addressing pollinator risks, their enforcement frequently remains inadequate. This implementation gap primarily arises from two key limitations: insufficient on-the-ground monitoring mechanisms and penalties that lack deterrence for non-compliance [169]. For example, systemic pesticides, which are known to have adverse effects on pollinator health, remain in use in regions where legal frameworks are fragmented or not stringently implemented. Divergent regulatory standards across nations and trade blocs further create conflicts that undermine efforts to mitigate pollinator risks. A case in point is that pesticides banned in the EU are frequently used in export-oriented agricultural systems elsewhere, exposing pollinator populations to transboundary hazards [182]. Furthermore, policies often fail to incentivize the adoption of IPM, which emphasizes non-chemical control measures as a primary strategy and chemical pesticides as a last resort. For instance, market forces and policy structures in major agricultural economies continue to favor chemical-intensive practices over biologically based pest control methods [183]. Additionally, policymaking processes frequently overlook key stakeholders, including smallholder farmers, local communities, and non-governmental organizations, who have a direct role in implementing pollinator-friendly practices. Without adequate involvement, resistance to "top–down" measures persists, reducing the success of new policies [184].

8. Research gaps and future directions

Despite significant advances in understanding how pesticides impact bee health, several critical research gaps remain. Addressing these gaps is vital for developing effective mitigation strategies and informing evidence-based policy.

8.1 Need for long-term, field-based studies

Most existing data are derived from short-term laboratory experiments that fail to replicate the complexity of real-world exposure. Bees in agricultural landscapes face continuous and multifactorial stressors, including seasonal pesticide applications, landscape fragmentation, and fluctuating floral resources

[6]. Long-term, field-based monitoring is necessary to assess chronic and transgenerational effects, such as reduced reproductive output or colony collapse over multiple generations [185]. These studies should integrate ecological variables, such as climate, crop type, and exposure timing.

8.2 In-depth elucidation of molecular mechanisms

While traditional ecotoxicology often relies on physiological, biochemical, and single molecular indicators to assess pollutant toxicity, this approach yields limited data for in-depth analysis of toxicity mechanisms. At the molecular level, many pesticides, particularly neonicotinoids, act on neural receptors such as nicotinic acetylcholine receptors (nAChRs), leading to neural overstimulation, oxidative stress, and apoptosis [186]. However, detailed pathways linking sublethal exposure to neurodevelopmental or behavioral impairments remain poorly understood. Genomic and transcriptomic studies have revealed altered expression of genes involved in neurotransmission, detoxification, and immune response, suggesting broader physiological consequences than initially assumed [187].

8.3 Integration of omics technologies in toxicology

Omics technologies offer powerful tools to characterize complex pesticide responses. Metabolomics and proteomics can identify biomarkers of sublethal stress, while transcriptomic profiling can reveal gene networks disrupted by pesticide exposure [188, 189]. These approaches enhance mechanistic understanding and support risk assessment by providing early warning indicators before overt colony-level symptoms emerge.

8.4 Unresolved questions and emerging technologies

Fundamental questions remain regarding the delayed and cumulative effects of low-dose exposure, synergistic interactions with pathogens, and the ecological consequences of pesticide mixtures [190]. New technologies, such as gene editing (e.g., CRISPR), may allow for the development of more resilient bee strains, although ecological risks and ethical concerns require thorough consideration [191].

8.5 Cross-disciplinary and translational research

Future research should integrate expertise from toxicology, molecular biology, ecology, agronomy, and policy studies. Translational efforts are needed to convert scientific findings into farmer-oriented practices and evidence-based regulations. This will require partnerships between academic researchers, regulators, NGOs, and agricultural stakeholders.

8.6 Multi-species and ecosystem-level approaches

The overwhelming focus on *Apis mellifera* limits ecological generalizability. Bumblebees (*Bombus* spp.), solitary bees (e.g., *Osmia* spp.), and other wild pollinators may exhibit greater sensitivity due to differences in life history, body size, or detoxification pathways [8, 192]. Research must expand to include these species and investigate pesticide effects at the level of pollinator communities, plant–pollinator networks, and ecosystem services [193].

9. Conclusion

A growing body of evidence demonstrates that pesticides can impair bee growth, development, and behavior through both acute and chronic mechanisms. These impacts manifest across biological levels, from molecular disruption and impaired neurodevelopment to colony-level dysfunction and population declines [13, 194]. Given the central role of bees in ecosystem functioning and food security, such threats are a cause for urgent global concern.

Robust regulatory reform is needed to protect pollinators. Risk assessment frameworks should incorporate sublethal, behavioral, and synergistic endpoints, moving beyond LD_{50} values [195]. The EU's precautionary ban on neonicotinoids serves as a model for science-based regulatory action. National governments should establish and maintain open-access bee toxicity databases to consolidate global knowledge on species-specific risks and pesticide interactions. In parallel, training programs for farmers and pesticide applicators must be expanded to promote safe application practices and awareness of pollinator risks. Monitoring and decision-support tools, such as biosensor-based pesticide residue detectors and exposure models such as BeeREX, should be deployed more widely [196]. Such tools enable farmers to optimize pesticide timing and minimize off-target exposure.

Moving forward, research and policy must converge to address critical knowledge gaps, particularly in areas such as sublethal effects on bee physiology and behavior, long-term population-level studies across diverse species, and interactions with stressors like climate change. These gaps hinder the development of comprehensive protection strategies. Implementing pollinator-safe agricultural systems requires filling these voids through targeted studies. Promising innovations, such as biopesticides, microbial formulations, and dose-optimized delivery systems (e.g., nanoparticle carriers), offer environmentally benign alternatives. Selective breeding or gene-editing approaches may enhance resistance traits in bees, provided they are ethically and ecologically justified.

Conflict of Interest

The authors declare no conflicts of interest.

Author details

Lifeng Meng[1]*, Jia Xing Huang[1] and Li Jiao Gao[2]

1 State Key Laboratory of Resource Insects, Key Laboratory for Insect-Pollinator Biology of the Ministry of Agriculture and Rural Affairs, Institute of Apicultural Research, Chinese Academy of Agricultural Sciences, Beijing, China

2 Department of Economic Animal, Chongqing Academy of Animal Sciences, Chongqing, China

*Address all correspondence to: menglifeng@caas.cn

References

[1] Klein A-M, Vaissière BE, Cane JH, et al. Importance of pollinators in changing landscapes for world crops. Proceedings of the Royal Society B: Biological Sciences. 2006;**274** (1608):303–313. DOI: 10.1098/rspb.2006.3721

[2] Potts SG, Biesmeijer JC, Kremen C, et al. Global pollinator declines: Trends, impacts and drivers. Trends in Ecology & Evolution. 2010;**25**(6):345–353. DOI: 10.1016/j.tree.2010.01.007

[3] Faichnie R, Breeze TD, Senapathi D, et al. Scales matter: Maximising the effectiveness of interventions for pollinators and pollination. The Future of Agricultural Landscapes, Part II. 2021:105–147. DOI: 10.1016/bs.aecr.2020.11.003

[4] Goulson D, Nicholls E, Botías C, et al. Bee declines driven by combined stress from parasites, pesticides, and lack of flowers. Science. 2015;**347** (6229). DOI: 10.1126/science.1255957

[5] Ward LT, Hladik ML, Guzman A, et al. Pesticide exposure of wild bees and honey bees foraging from field border flowers in intensively managed Agriculture areas. Science of the Total Environment. 2022:831. DOI: 10.1016/j.scitotenv.2022.154697

[6] Guedes RNC, Sanchez-Bayo F, Goka K. Pesticide residues and bees – A risk assessment. PLoS One. 2014;**9**(4). DOI: 10.1371/journal.pone.0094482

[7] Lima YS, de Castro Lippi IC, da Luz Scheffer J, et al. Food contamination with fipronil alters gene expression associated with foraging in Africanized honey bees. Environmental Science and Pollution Research. 2024;**31** (39):52267–52278. DOI: 10.1007/s11356-024-34695-8

[8] Fisher II A, DeGrandi-Hoffman G, Liao LH, et al. The challenge of balancing fungicide use and pollinator health. In: Harrison JF, editor. Environmental Threats to Pollinator Health and Fitness. Cambridge, Massachusetts: Academic Press Inc; 2023. p. 117–190. DOI: 10.1016/bs.aiip.2023.01.002

[9] Dai P, Yan Z, Ma S, et al. The herbicide glyphosate negatively affects midgut bacterial communities and survival of honey bee during larvae reared in vitro. Journal of Agricultural and Food Chemistry. 2018;**66**(29):7786–7793. DOI: 10.1021/acs.jafc.8b02212

[10] Pilling ED, Jepson PC. Synergism between EBI fungicides and a pyrethroid insecticide in the honeybee (Apis mellifera). Pesticide Science. 1993;**39**(4):293–297. DOI: 10.1002/ps.2780390407

[11] Tsvetkov N, Samson-Robert O, Sood K, et al. Chronic exposure to neonicotinoids reduces honey bee health near corn crops. Science. 2017;**356**:1395–1397

[12] Ahsan Z, Wu Z, Lin Z, et al. The sublethal effects of neonicotinoids on honey bees. Biology. 2025;**14**(8):1076. DOI: 10.3390/biology14081076

[13] Gill RJ, Ramos-Rodriguez O, Raine NE. Combined pesticide exposure severely affects individual- and colony-level traits in bees. Nature.

2012;**491**(7422):105–108. DOI: 10.1038/nature11585

[14] Marion-Poll F, Wu JY, Anelli CM, et al. Sub-lethal effects of pesticide residues in brood comb on worker honey bee (Apis mellifera) development and longevity. PLoS One. 2011;**6**(2). DOI: 10.1371/journal.pone.0014720

[15] Urueña Á, Blasco-Lavilla N, De la Rúa P. Sulfoxaflor effects depend on the interaction with other pesticides and Nosema ceranae infection in the honey bee (Apis mellifera). Ecotoxicology and Environmental Safety. 2023;264. DOI: 10.1016/j.ecoenv.2023.115427

[16] Van der Sluijs JP, Simon-Delso N, Goulson D, et al. Neonicotinoids, bee disorders and the sustainability of pollinator services. Current Opinion in Environmental Sustainability. 2013;**5**(3-4):293–305. DOI: 10.1016/j.cosust.2013.05.007

[17] Shu-dong LUO, Jian-dong AN, Ji-lian LI, et al. Hazard assessment and methods to measure the acute toxicity of chemical pesticides to bees. Journal of Hunan Agricultural University (Natural Sciences). 2009;**35**(3):320–324

[18] Farruggia FT, Garber K, Hartless C, et al. A retrospective analysis of honey bee (Apis mellifera) pesticide toxicity data. PLoS One. 2022;**17**(4):e0265962. DOI: 10.1371/journal.pone.0265962

[19] Li B, Ke L, Li A-R, et al. Exposure of larvae to sublethal thiacloprid delays bee development and affects transcriptional responses of newly emerged honey bees. Frontiers in Insect Science. 2022:2. DOI: 10.3389/finsc.2022.844957

[20] Williams GR, Troxler A, Retschnig G, et al. Neonicotinoid pesticides severely affect honey bee queens. Scientific Reports. 2015;**5**(1). DOI: 10.1038/srep14621

[21] Dai P, Jack CJ, Mortensen AN, et al. Acute toxicity of five pesticides to Apis mellifera larvae reared in vitro. Pest Management Science. 2017;**73**(11):2282–2286. DOI: 10.1002/ps.4608

[22] O'Reilly AO, Williamson MS, González-Cabrera J, et al. Predictive 3D modelling of the interactions of pyrethroids with the voltage-gated sodium channels of ticks and mites. Pest Management Science. 2013;**70**(3):369–377. DOI: 10.1002/ps.3561

[23] Heller S, Fine J, Phan NT, et al. Toxicity of formulated systemic insecticides used in apple orchard pest management programs to the honey bee (Apis mellifera L.). Environments. 2022;**9**(7):90. DOI: 10.3390/environments9070090

[24] Stoner KA, Eitzer BD. Using a hazard quotient to evaluate pesticide residues detected in pollen trapped from honey bees (Apis mellifera) in Connecticut. PLoS One. 2013;**8**(10):e77550. DOI:10.1371/journal.pone.0077550

[25] Kniss AR, De Stefano A, Arnold E, et al. Honeybee toxicity of pesticides used in United States maize and soybean production, 1998–2020. Integrated Environmental Assessment and Management. 2025;**21**(3):604–613. DOI: 10.1093/inteam/vjaf003

[26] Almasri H, Tavares DA, Pioz M, Sené D, et al. Mixtures of an insecticide, a fungicide and a herbicide induce high toxicities and systemic physiological disturbances in winter Apis mellifera honey bees. Ecotoxicology and Environmental Safety. 2020:203. DOI: 10.1016/j.ecoenv.2020.111013

[27] Végh R, Csóka M, Mednyánszky Z, et al. Pesticide residues in bee bread, propolis, beeswax and royal jelly – A review of the literature and dietary risk assessment. Food and Chemical Toxicology. 2023:176. DOI: 10.1016/j.fct.2023.113806

[28] Al-Waili N, Salom K, Al-Ghamdi A, et al. Antibiotic, pesticide, and microbial contaminants of honey: Human health hazards. The Scientific World Journal. 2012;**2012**:1–9

[29] Kast C, Kilchenmann V. An in vitro model for assessing the toxicity of pesticides in beeswax on honey bee larvae. Chemosphere. 2022;287. DOI: 10.1016/j.chemosphere.2021.132214

[30] Castano-Sanz V, Gomez-Mestre I, Garcia-Gonzalez F. Evolutionary consequences of pesticide exposure include transgenerational plasticity and potential terminal investment transgenerational effects. Evolution. 2022;**76**(11):2649–2668. DOI: 10.1111/evo.14613

[31] Calatayud-Vernich P, Calatayud F, Simó E, et al. Pesticide residues in honey bees, pollen and beeswax: Assessing beehive exposure. Environmental Pollution. 2018;**241**:106–114

[32] Bonzini S, Tremolada P, Bernardinelli I, et al. Predicting pesticide fate in the hive (part 1): Experimentally determined τ-fluvalinate residues in bees, honey and wax. Apidologie. 2011;**42**(3):378–390. DOI: 10.1007/s13592-011-0011-2

[33] Günes ME, Sari MF, Esen F. Organochlorine pesticides in honeybee, pollen and honey in bursa, Turkey. Food Additives & Contaminants: Part B. 2021;**14**(2):126–132. DOI: 10.1080/19393210.2021.1896583

[34] Sabo R, Staroň M, Sabová L, et al. Honey bees for pesticide monitoring in the landscape: Which bee matrices should be used? Chemosphere. 2024;**364**:143130. 143130. DOI: 10.1016/j.chemosphere.2024.143130

[35] Guo H, Xiao N, Liang Y, et al. Neurotoxic impacts of neonicotinoid pesticide residues in Chinese honey on Apis mellifera L. Journal of the Science of Food and Agriculture. 2025; DOI: 10.1002/jsfa.70154

[36] Manning R. Chemical residues in beebread, honey, pollen and wax samples collected from bee hives placed on canola crops in western Australia. Journal of Apicultural Research. 2018;**57**(5):696–708. DOI: 10.1080/00218839.2018.1494889

[37] Bischoff K, Baert N, McArt S. Pesticide contamination of beeswax from managed honey bee colonies in New York state. Journal of Veterinary Diagnostic Investigation. 2023;**35**(6):617–624. DOI: 10.1177/10406387231199098

[38] Benito-Murcia M, García-Vicente E, Martín M, et al. Pesticide residues in beeswax from western Spain: Toxicological risk for bee health and beekeeping productivity. Environmental Toxicology and Pharmacology. 2025;**117**:104744. 104744. DOI: 10.1016/j.etap.2025.104744

[39] Mullin CA, Frazier M, Frazier JL, et al. High levels of miticides and agrochemicals in North American apiaries: Implications for honey bee health. PLoS One. 2010;**5**(3):e9754. DOI: 10.1371/journal.pone.0009754

[40] Chowdhury AR, Gupta S. An overview of complex threats to bee populations: Pesticides, contaminants, and the urgency for research. Proceedings of the National Academy of Sciences, India Section B: Biological Sciences. 2025. DOI: 10.1007/s40011-0 25-01741-5

[41] Belsky J, Joshi NK. Impact of biotic and abiotic stressors on managed and feral bees. Insects. 2019;**10**(8):233. DOI: 10.3390/insects10080233

[42] Saleem MS, Huang ZY, Milbrath MO. Neonicotinoid pesticides are more toxic to honey bees at lower temperatures: Implications for overwintering bees. Frontiers in Ecology and Evolution. 2020;8. DOI: 10.3389/fevo.2020.556856

[43] Mallinger RE, Werts P, Gratton C. Pesticide use within a pollinator-dependent crop has negative effects on the abundance and species richness of sweat bees, Lasioglossum spp., and on bumble bee colony growth. Journal of Insect Conservation. 2015;**19**(5):999–1010. DOI: 10.1007/s10841-015-9816-z

[44] St. Clair AL, Zhang G, Dolezal AG, et al. Diversified farming in a monoculture landscape: Effects on honey bee health and wild bee communities. Environmental Entomology. 2020;**49**(3):753–764. DOI: 10.1093/ee/nvaa031

[45] Siviter H, Bailes EJ, Martin CD, et al. Agrochemicals interact synergistically to increase bee mortality. Nature. 2021;**596**(7872):389–392. DOI: 10.1038/s41586-021-03787-7

[46] Wang Y, Zhu Y-C, Li W, et al. Binary and ternary toxicological interactions of clothianidin and eight commonly used pesticides on honey bees (Apis mellifera). Ecotoxicology and Environmental Safety. 2021;**223**:112563

[47] Raine NE, Johnson RM, Dahlgren L, et al. Acaricide, fungicide and drug interactions in honey bees (Apis mellifera). PLoS One. 2013;**8**(1). DOI: 10.1371/journal.pone.0054092

[48] Annoscia D, Di Prisco G, Becchimanzi A, et al. Neonicotinoid clothianidin reduces honey bee immune response and contributes to varroa mite proliferation. Nature Communications. 2020;**11**(1). DOI: 10.1038/s41467-020-19715-8

[49] Brandt A, Gorenflo A, Siede R, et al. The neonicotinoids thiacloprid, imidacloprid, and clothianidin affect the immunocompetence of honey bees (Apis mellifera L.). Journal of Insect Physiology. 2016;**86**:40–47

[50] Alaux C, Brunet JL, Dussaubat C, et al. Interactions between Nosema microspores and a neonicotinoid weaken honeybees (Apis mellifera). Environmental Microbiology. 2010;**12** (3):774–782. DOI: 10.1111/j.1462-2920.2009.02123.x

[51] Paris L, Peghaire E, Moné A, et al. Honeybee gut microbiota dysbiosis in pesticide/parasite co-exposures is mainly induced by Nosema ceranae. Journal of Invertebrate Pathology. 2020:172. DOI: 10.1016/j.jip.2020.107348

[52] Doublet V, Labarussias M, de Miranda JR, et al. Bees under stress: Sublethal doses of a neonicotinoid pesticide and pathogens interact to elevate honey bee mortality across the

life cycle. Environmental Microbiology. 2014;**17**(4):969–983. DOI: 10.1111/1462-2920.12426

[53] Morfin N, Goodwin PH, Hunt GJ, et al. Effects of sublethal doses of clothianidin and/or v. Destructor on honey bee (Apis mellifera) self-grooming behavior and associated gene expression. Scientific Reports. 2019;**9**(1). DOI: 10.1038/s41598-019-41365-0

[54] Zeil J, Di Pasquale G, Salignon M, et al. Influence of pollen nutrition on honey bee health: Do pollen quality and diversity matter? PLoS One. 2013;**8**(8). DOI: 10.1371/journal.pone.0072016

[55] Almasri H, Tavares DA, Diogon M, et al. Physiological effects of the interaction between Nosema ceranae and sequential and overlapping exposure to glyphosate and difenoconazole in the honey bee Apis mellifera. Ecotoxicology and Environmental Safety. 2021;**217**:112258

[56] Wang Y, Li Z, Qin Q, et al. Imidacloprid Disrupts Larval Molting Regulation and Nutrient Energy Metabolism, Causing Developmental Delay in Honey Bee Apis Mellifera. 2024. 12 p. DOI: 10.7554/eLife.88772

[57] Friol PS, Catae AF, Tavares DA, et al. Can the exposure of Apis mellifera (Hymenoptera, Apidae) larvae to a field concentration of thiamethoxam affect newly emerged bees? Chemosphere. 2017;**185**:56–66

[58] Kopit AM, Klinger E, Cox-Foster DL, et al. Effects of Provision Type and Pesticide Exposure on the Larval Development of Osmia lignaria (Hymenoptera: Megachilidae). Environmental Entomology. 2021;**51**(1):240–251. DOI: 10.1093/ee/nvab119

[59] Reddy GVP, Vázquez DE, Ilina N, et al. Glyphosate affects the larval development of honey bees depending on the susceptibility of colonies. PLoS One. 2018;**13**(10). DOI: 10.1371/journal.pone.0205074

[60] Vázquez DE, Latorre-Estivalis JM, Ons S, et al. Chronic exposure to glyphosate induces transcriptional changes in honey bee larva: A toxicogenomic study. Environmental Pollution. 2020:261. DOI: 10.1016/j.envpol.2020.114148

[61] Dai P, Jack CJ, Mortensen AN, et al. Chronic toxicity of amitraz, coumaphos and fluvalinate to Apis mellifera L. Larvae reared in vitro. Scientific Reports. 2018;**8**(1). DOI: 10.1038/s41598-018-24045-3

[62] Cook SC. Compound and dose-dependent effects of two neonicotinoid pesticides on honey bee (Apis mellifera) metabolic physiology. Insects. 2019;**10**(1):18. DOI:10.3390/insects10010018

[63] Bruckner S, Straub L, Villamar-Bouza L, et al. Life stage dependent effects of neonicotinoid exposure on honey bee hypopharyngeal gland development. Ecotoxicology and Environmental Safety. 2024:288. DOI: 10.1016/j.ecoenv.2024.117337

[64] Arthidoro de Castro MB, Martinez LC, Cossolin JFS, et al. Cytotoxic effects on the midgut, hypopharyngeal, glands and brain of Apis mellifera honey bee workers exposed to chronic concentrations of lambda-cyhalothrin. Chemosphere. 2020:248. DOI: 10.1016/j.chemosphere.2020.126075

[65] Heylen K, Gobin B, Arckens L, et al. The effects of four crop

protection products on the morphology and ultrastructure of the hypopharyngeal gland of the European honeybee. Apis Mellifera. Apidologie. 2011;**42**(1):103–116. DOI: 10.1051/apido/2010043

[66] Menail AH, Bouchema WF, Mansouri Z, et al. 2015 Negative Sublethal Effect of the Insecticide Thiamethoxam on Honeybees. In: 12th International Symposium of the ICP-PR Bee Protection Group - Hazards of Pesticides to Bees. Ghent Univ, Fac Bioscience Engn, Ghent, Belgium: Julius-Kühn-Archiv; SEP 15–17, 2014; **450**:151–151

[67] Cartereau A, Martin C, Thany SH. Neonicotinoid insecticides differently modulate acetylcholine-induced currents on mammalian α7 nicotinic acetylcholine receptors. British Journal of Pharmacology. 2017;**175**(11): 1987–1998. DOI: 10.1111/bph.14018

[68] Fine JD, Corby-Harris V. Beyond brood: The potential impacts of insect growth disruptors on the long-term health and performance of honey bee colonies. Apidologie. 2021;**52** (3):580–595. DOI: 10.1007/s13592-021-00845-x

[69] Christen V, Schirrmann M, Frey JE, et al. Global Transcriptomic Effects of Environmentally Relevant Concentrations of the Neonicotinoids Clothianidin, Imidacloprid, and Thiamethoxam in the Brain of Honey Bees (Apis mellifera). Environmental Science & Technology. 2018;**52** (13):7534–7544. DOI: 10.1021/acs.est.8b01801

[70] Oliveira MSD, Pereira GDS, Martinez LC, et al. Effects of chronic oral exposure to insecticide teflubenzuron on the midgut of the honey bee Apis mellifera workers: Histopathological insights into pesticide toxicity. Environmental Science and Pollution Research. 2024;**31**(32): 44908–44919. DOI: 10.1007/s11356-024-34066-3

[71] Reis AB, Martínez LC, de Oliveira MS, et al. Sublethal Effects Induced by a Cyflumetofen Formulation on Honeybee Apis mellifera L. Workers: Assessment of Midgut, Hypopharyngeal Glands, and Fat Body Integrity. Environmental Toxicology and Chemistry. 2024;**43**(11):2455–2465. DOI: 10.1002/etc.5980

[72] Paoli M, Giurfa M. Pesticides and pollinator brain: How do neonicotinoids affect the central nervous system of bees? European Journal of Neuroscience. 2024;**60**(8):5927–5948. DOI: 10.1111/ejn.16536

[73] Tavares DA, Roat TC, Silva-Zacarin ECM, et al. Exposure to thiamethoxam during the larval phase affects synapsin levels in the brain of the honey bee. Ecotoxicology and Environmental Safety. 2019;**169**: 523–528

[74] Peng Y-C, Yang E-C. Sublethal dosage of imidacloprid reduces the microglomerular density of honey bee mushroom bodies. Scientific Reports. 2016;**6**(1). DOI: 10.1038/srep19298

[75] Williamson SM, Wright GA. Exposure to multiple cholinergic pesticides impairs olfactory learning and memory in honeybees. Journal of Experimental Biology. 2013. DOI: 10.1242/jeb.083931

[76] Tsvetkov N, Moon KM, Foster LJ. Proteome of clothianidin exposed honey bees reveals a possible mechanism behind impairment of sucrose

responsiveness. Journal of Applied Entomology. 2024;**149**(1):121–131. DOI: 10.1111/jen.13363

[77] Tison L, Rößner A, Gerschewski S, et al. The neonicotinoid clothianidin impairs memory processing in honey bees. Ecotoxicology and Environmental Safety. 2019;**180**:139–145

[78] Henry M, Béguin M, Requier F, et al. A common pesticide decreases foraging success and survival in honey bees. Science. 2012;**336**:348–350

[79] Fent K, Haltiner T, Kunz P, et al. Insecticides cause transcriptional alterations of endocrine related genes in the brain of honey bee foragers. Chemosphere. 2020:260. DOI: 10.1016/j.chemosphere.2020.127542

[80] Abou-Shaara HF. The response of heat shock proteins in honey bees to abiotic and biotic stressors. Journal of Thermal Biology. 2024;119. DOI: 10.1016/j.jtherbio.2024.103784

[81] McKinstry M, Prado-Irwin SR, Adames TR, et al. Retained metabolic activity in honey bee collected pollen has implications for pollen digestion and effects on honey bee health. Apidologie. 2020;**51**(2):212–225. DOI: 10.1007/s13592-019-00703-x

[82] Yao J, Zhu YC, Adamczyk J, et al. Influences of acephate and mixtures with other commonly used pesticides on honey bee (Apis mellifera) survival and detoxification enzyme activities. Comparative Biochemistry and Physiology Part C: Toxicology & Pharmacology. 2018;**209**:9–17

[83] Quinlan G, Döke MA, Ortiz-Alvarado Y, et al. Carbohydrate nutrition associated with health of overwintering honey bees. Journal of Insect Science. 2023;**23**(6). DOI: 10.1093/jisesa/iead084

[84] Soydan E, Olcay AC, Bilir G, et al. Investigation of pesticides on honey bee carbonic anhydrase inhibition. Journal of Enzyme Inhibition and Medicinal Chemistry. 2020;**35**(1):1923–1927. DOI: 10.1080/14756366.2020.1835885

[85] Motta EVS, Raymann K, Moran NA. Glyphosate perturbs the gut microbiota of honey bees. Proceedings of the National Academy of Sciences. 2018;**115**(41):10305–10310. DOI: 10.1073/pnas.1803880115

[86] Souza DAD, Huang MH, Tarpy DR. Experimental improvement of honey bee (Apis mellifera) queen quality through nutritional and hormonal supplementation. Apidologie. 2018. DOI: 10.1007/s13592-018-0614-y

[87] Fine JD. Evaluation and comparison of the effects of three insect growth regulators on honey bee queen oviposition and egg eclosion. Ecotoxicology and Environmental Safety. 2020;205. DOI: 10.1016/j.ecoenv.2020.111142

[88] Pineaux M, Grateau S, Lirand T, et al. Honeybee queen exposure to a widely used fungicide disrupts reproduction and colony dynamic. Environmental Pollution. 2023:322. DOI: 10.1016/j.envpol.2023.121131

[89] Dussaubat C, Maisonnasse A, Crauser D, et al. Combined neonicotinoid pesticide and parasite stress alter honeybee queens' physiology and survival. Scientific Reports. 2016;**6**(1). DOI: 10.1038/srep31430

[90] Walsh EM, Sweet S, Knap A, et al. Queen honey bee (Apis mellifera)

pheromone and reproductive behavior are affected by pesticide exposure during development. Behavioral Ecology and Sociobiology. 2020;**74**(3). DOI: 10.1007/s00265-020-2810-9

[91] Walsh EM, Janowiecki MA, Zhu K, et al. Elevated mating frequency in honey bee (Hymenoptera: Apidae) queens exposed to the miticide amitraz during development. Annals of the Entomological Society of America. 2021;**114**(5):620–626. DOI: 10.1093/aes a/saaa041

[92] Desneux N, Fisher A, Rangel J. Exposure to pesticides during development negatively affects honey bee (Apis mellifera) drone sperm viability. PLoS One. 2018;**13** (12). DOI: 10.1371/journal. pone.0208630

[93] Johnson RM, Dahlgren L, Siegfried BD, et al. Effect of in-hive miticides on drone honey bee survival and sperm viability. Journal of Apicultural Research. 2015;**52**(2):88–95. DOI: 10.3896/ibra.1.52.2.18

[94] Ma C, Shi X, Chen S, et al. Combined pesticides in field doses weaken honey bee (Apis cerana F.) flight ability and analyses of transcriptomics and metabolomics. Pesticide Biochemistry and Physiology. 2024:201. DOI: 10.1016/j. pestbp.2024.105793

[95] Kairo G, Provost B, Tchamitchian S, et al. Drone exposure to the systemic insecticide fipronil indirectly impairs queen reproductive potential. Scientific Reports. 2016;**6**(1). DOI: 10.1038/srep31904

[96] Brown J, vanEngelsdorp D, Evans JD, et al. Colony collapse disorder: A descriptive study. PLoS One. 2009;**4**(8). DOI: 10.1371/journal. pone.0006481

[97] Tokach R, Smart A, Fassbinder-Orth C, et al. Honey bee colony behavior and ontogeny are adversely affected when exposed to a pesticide-contaminated environment. Journal of Insect Science. 2024;**24**(3). DOI: 10.1093/jisesa/ieae034

[98] Yang EC, Chuang YC, Chen YL, et al. Abnormal foraging behavior induced by sublethal dosage of imidacloprid in the honey bee (Hymenoptera: Apidae). Journal of Economic Entomology. 2008;**101** (6):1743–1748. DOI: 10.1603/0022-0493-101.6.1743

[99] Levitan IB, Kaczmarek LK. Navigation: Cognition, learning, and memory. Elsevier eBooks. 2024. DOI: 10.0410/cata/ f9c150e1a58161af76da983b1ae351da

[100] Sharma R, Arya S, Singh R. Pesticides herbicides and their effects in pollination. International Journal of Environmental Science. 2025;**16** (02):110–119. DOI: 10.53390/ ijes.2025.16204

[101] Tosi S, Burgio G, Nieh JC. A common neonicotinoid pesticide, thiamethoxam, impairs honey bee flight ability. Scientific Reports. 2017;**7**(1). DOI: 10.1038/s41598-017-01361-8

[102] Siviter H, Koricheva J, Brown MJF, et al. Quantifying the impact of pesticides on learning and memory in bees. Journal of Applied Ecology. 2018;**55**(6):2812–2821. DOI: 10.1111/1365-2664.13193

[103] Birkenbach M, Straub F, Kiesel A, et al. Land-use affects pollinator-specific resource availability and pollinator

foraging behaviour. Ecology and Evolution. 2024;**14**(3). DOI: 10.1002/ece3.11061

[104] Kessler SC, Tiedeken EJ, Simcock KL, et al. Bees prefer foods containing neonicotinoid pesticides. Nature. 2015;**521**(7550):74–76. DOI: 10.1038/nature14414

[105] Eiri DM, Nieh JC. A nicotinic acetylcholine receptor agonist affects honey bee sucrose responsiveness and decreases waggle dancing. Journal of Experimental Biology. 2012;**215**(12):2022–2029. DOI: 10.1242/jeb.068718

[106] Wainselboim AJ, Farina WM. Trophallaxis in the honeybee Apis mellifera (L.): The interaction between flow of solution and sucrose concentration of the exploited food sources. Animal Behaviour. 2000;**59**(6):1177–1185. DOI: 10.1006/anbe.1999.1418

[107] Encerrado-Manriquez AM, Pouv AK, Fine JD, et al. Enhancing knowledge of chemical exposures and fate in honey bee hives: Insights from colony structure and interactions. Science of the Total Environment. 2024:916. DOI: 10.1016/j.scitotenv.2024.170193

[108] Williamson SM, Willis SJ, Wright GA. Exposure to neonicotinoids influences the motor function of adult worker honeybees. Ecotoxicology. 2014;**23**(8):1409–1418. DOI: 10.1007/s10646-014-1283-x

[109] Chakrabarti P, Rana S, Bandopadhyay S, et al. Field populations of native Indian honey bees from pesticide intensive agricultural landscape show signs of impaired olfaction. Scientific Reports. 2015;**5**(1). DOI: 10.1038/srep12504

[110] Spivak M, Danka RG. Perspectives on hygienic behavior in Apis mellifera and other social insects. Apidologie. 2020;**52**(1):1–16. DOI: 10.1007/s13592-020-00784-z

[111] Erez T, Bonda E, Kahanov P, et al. Multiple benefits of breeding honey bees for hygienic behavior. Journal of Invertebrate Pathology. 2022:193. DOI: 10.1016/j.jip.2022.107788

[112] Schuehly W, Riessberger-Gallé U, Hernández López J. Sublethal pesticide exposure induces larval removal behavior in honeybees through chemical cues. Ecotoxicology and Environmental Safety. 2021;228. DOI: 10.1016/j.ecoenv.2021.113020

[113] Olivares-Pinto U, Alaux C, Le Conte Y, et al. Using honey bee flight activity data and a deep learning model as a toxicovigilance tool. Ecological Informatics. 2024:81. DOI: 10.1016/j.ecoinf.2024.102653

[114] Adamczyk S, Lázaro R, Pérez-Arquillué C, et al. Impact of the use of fluvalinate on different types of beeswax from Spanish hives. Archives of Environmental Contamination and Toxicology. 2009;**58**(3):733–739. DOI: 10.1007/s00244-009-9387-7

[115] Morfin N, Fillier TA, Pham TH, et al. First insights into the honey bee (Apis mellifera) brain lipidome and its neonicotinoid-induced alterations associated with reduced self-grooming behavior. Journal of Advanced Research. 2022;**37**:75–89

[116] Sakamoto Y, Yoshiyama M, Maeda T, et al. Effects of neonicotinoids

on honey bee autogrooming behavior against the tracheal mite Acarapis woodi. Ecotoxicology. 2022;**31**(2): 251–258. DOI: 10.1007/s10646-021-02503-5

[117] Foose AM, Westwick RR, Vengarai M, et al. The survival consequences of grooming in the honey bee Apis mellifera. Insectes Sociaux. 2022;**69**(2-3):279–287. DOI: 10.1007/s00040-022-00868-2

[118] Qureshi A, Mohan M, Kanade GS, et al. In situ bioremediation of organochlorine-pesticide-contaminated microcosm soil and evaluation by gene probe. Pest Management Science. 2009;**65**(7):798–804. DOI: 10.1002/ps.1757

[119] Blacquière T, Smagghe G, van Gestel CAM, et al. Neonicotinoids in bees: A review on concentrations, side-effects and risk assessment. Ecotoxicology. 2012;**21**(4):973–992. DOI: 10.1007/s10646-012-0863-x

[120] EFSA, Neonicotinoids: Risks to bees confirmed. European food safety authority. 2018. https://www.efsa.europa.eu/en/press/news/180228 [Accessed: 2025-August-06]

[121] Brown MJF, Paxton RJ. The conservation of bees: A global perspective. Apidologie. 2009;**40**(3):410–416. DOI: 10.1051/apido/2009019

[122] Palmer MJ, Moffat C, Saranzewa N, et al. Cholinergic pesticides cause mushroom body neuronal inactivation in honeybees. Nature Communications. 2013;**4**(1). DOI: 10.1038/ncomms2648

[123] Naug D, Fischer J, Müller T, et al. Neonicotinoids interfere with specific components of navigation in honeybees. PLoS One. 2014;**9**(3). DOI: 10.1371/journal.pone.0091364

[124] Catae AF, Roat TC, Pratavieira M, et al. Exposure to a sublethal concentration of imidacloprid and the side effects on target and nontarget organs of Apis mellifera (Hymenoptera, Apidae). Ecotoxicology. 2017;**27**(2):109–121. DOI: 10.1007/s10646-017-1874-4

[125] Li Z, Yu T, Chen Y, et al. Brain transcriptome of honey bees (Apis mellifera) exhibiting impaired olfactory learning induced by a sublethal dose of imidacloprid. Pesticide Biochemistry and Physiology. 2019;**156**:36–43

[126] Benjamin SG, Nandini NJ. Effect of endocrine disruptor pesticides on nontarget pollinator-honey bee. Sustainable Agriculture, Food and Environmental Research. 2023;11. DOI: 10.7770/safer.v11i1.2823

[127] Milone JP, Tarpy DR. Effects of developmental exposure to pesticides in wax and pollen on honey bee (Apis mellifera) queen reproductive phenotypes. Scientific Reports. 2021;**11**(1). DOI: 10.1038/s41598-020-80446-3

[128] Mansour R, Fine JD, Foster LJ, et al. Indirect exposure to insect growth disruptors affects honey bee (Apis mellifera) reproductive behaviors and ovarian protein expression. PLoS One. 2023;**18**(10). DOI: 10.1371/journal.pone.0292176

[129] Litsey EM, Fine JD, Tarpy D. Developmental exposure to hormone-mimicking insect growth disruptors alters expression of endocrine-related genes in worker honey bee (Hymenoptera: Apidae) brains and hypopharyngeal glands.

Journal of Economic Entomology. 2024;**117**(2):377–387. DOI: 10.1093/jee/toae006

[130] Parrott JL, Metcalfe CD. Assessing the effects of the antidepressant venlafaxine to fathead minnows exposed to environmentally relevant concentrations over a full life cycle. Environmental Pollution. 2017;**229**:403–411. DOI: 10.1016/j.envpol.2017.06.009

[131] Gong Y, Diao Q. Current knowledge of detoxification mechanisms of xenobiotic in honey bees. Ecotoxicology. 2016;**26**(1):1–12. DOI: 10.1007/s10646-016-1742-7

[132] Stoner KA. Current pesticide risk assessment protocols do not adequately address differences between honey bees (Apis mellifera) and bumble bees (bombus spp.). Frontiers in Environmental Science. 2016;4. DOI: 10.3389/fenvs.2016.00079

[133] Thompson HM, Hunt LV. Extrapolating from honeybees to bumblebees in pesticide risk assessment. Ecotoxicology. 1999;**8**(3):147–166. DOI: 10.1023/a:1026444029579

[134] Gill RJ, Raine NE, Carroll S. Chronic impairment of bumblebee natural foraging behaviour induced by sublethal pesticide exposure. Functional Ecology. 2014;**28**(6):1459–1471. DOI: 10.1111/1365-2435.12292

[135] Zioga E, White B, Stout JC. Honey bees and bumble bees may be exposed to pesticides differently when foraging on agricultural areas. Science of the Total Environment. 2023;896. DOI: 10.1016/j.scitotenv.2023.166214

[136] Stanley DA, Russell AL, Morrison SJ, et al. Investigating the impacts of field-realistic exposure to a neonicotinoid pesticide on bumblebee foraging, homing ability and colony growth. Journal of Applied Ecology. 2016;**53**(5):1440–1449. DOI: 10.1111/1365-2664.12689

[137] Xu J, Strange JP, Welker DL, et al. Detoxification and stress response genes expressed in a western North American bumble bee, bombus huntii (Hymenoptera: Apidae). BMC Genomics. 2013;**14**(1). DOI: 10.1186/1471-2164-14-874

[138] Nasir M. Susceptibility comparison of bumblebee and honeybee to acetamiprid and imidacloprid. Jammu Kashmir Journal of Agriculture. 2023;**1**(1):01–06. DOI: 10.56810/jkjagri.001.01.0006

[139] Dinter A, Alix A, Becker R, et al. 2018 Non-apis (Bombus Terrestris) versus Honeybee (Apis Mellifera) Acute Oral and Contact Sensitivity – Preliminary Results of ECPA Company Data Evaluation. In: 13th International Symposium of the ICP-PR-Bee-Protection-Group on Hazards of Pesticides to Bees. Julius-Kühn-Archiv, Nr; OCT 18-20, 2017. **462**. Valencia, Spain, 170-172 p. DOI: 10.5073/JKA.2018.462.050

[140] Vassvik L, Nielsen A, Garratt MPD, et al. Solitary bees reduce pollination and production deficits in apple cultivation. Agriculture, Ecosystems & Environment. 2025:388. DOI: 10.1016/j.agee.2025.109670

[141] O'Reilly AD, Stanley DA. Solitary bee behaviour and pollination service delivery is differentially impacted by neonicotinoid and pyrethroid insecticides. Science of the Total Environment. 2023;894. DOI: 10.1016/j.scitotenv.2023.164399

[142] Luu C, Hageman KJ, Singer TL, et al. Pesticide transfer between solitary bee nesting materials and provisions—implications for larval exposure. Environmental Science & Technology. 2025;**59**(35):18904–18914. DOI: 10.1021/acs.est.5c08032

[143] Kopit AM, Pitts-Singer TL. Routes of pesticide exposure in solitary, cavity-nesting bees. Environmental Entomology. 2018;**47**(3):499–510. DOI: 10.1093/ee/nvy034

[144] Phana NT, Joshia NK, Rajottec EG, et al. Systemic pesticides in a solitary bee pollen food store affect larval development and increase pupal mortality. Science of the Total Environment. 2024:915. DOI: 10.1016/j.scitotenv.2024.170048

[145] Willis Chan DS, Raine NE. Neonicotinoid exposure affects foraging, nesting, and reproductive success of ground-nesting solitary bees. bioRxiv. 2020. DOI: 10.1101/2020.10.07.330605

[146] Boff S, Scheiner R, Raizer J, et al. Survival rate and changes in foraging performances of solitary bees exposed to a novel insecticide. Ecotoxicology and Environmental Safety. 2021;**211**:111869. DOI: 10.1016/j.ecoenv.2020.111869

[147] Martins CAH, Caliani I, D'Agostino A, et al. Biochemical responses, feeding and survival in the solitary bee osmia bicornis following exposure to an insecticide and a fungicide alone and in combination. Environmental Science and Pollution Research. 2022;**30**(10):27636–27649. DOI: 10.1007/s11356-022-24061-x

[148] Albacete S, Azpiazu C, Sancho G, et al. Sublethal fungicide-insecticide co-exposure affects nest recognition and parental investment in a solitary bee. Environmental Pollution. 2024;**363**:125223. DOI: 10.1016/j.envpol.2024.125223

[149] Knauer AC, Alaux C, Allan MJ, et al. Nutritional stress exacerbates impact of a novel insecticide on solitary bees' behaviour, reproduction and survival. Proceedings of the Royal Society B: Biological Sciences. 2022;**289**(1984). DOI: 10.1098/rspb.2022.1013

[150] Rondeau S, Raine NE. Single and combined exposure to 'bee safe' pesticides alter behaviour and offspring production in a ground-nesting solitary bee (Xenoglossa pruinosa). Proceedings of the Royal Society B: Biological Sciences. 2024;**291**(2019). DOI: 10.1098/rspb.2023.2939

[151] Hayward A, Beadle K, Singh KS, et al. The leafcutter bee, megachile rotundata, is more sensitive to n-cyanoamidine neonicotinoid and butenolide insecticides than other managed bees. Nature Ecology & Evolution. 2019;**3**(11):1521–1524. DOI: 10.1038/s41559-019-1011-2

[152] Beadle K, Singh KS, Troczka BJ, et al. Genomic insights into neonicotinoid sensitivity in the solitary bee Osmia bicornis. PLOS Genetics. 2019;**15**(2):e1007903. DOI: 10.1371/journal.pgen.1007903

[153] Hayward A, Hunt BJ, Haas J, et al. A cytochrome p450 insecticide detoxification mechanism is not conserved across the megachilidae family of bees. Evolutionary Applications. 2023;**17**(1). DOI: 10.1111/eva.13625

[154] Thompson HM, Pamminger T. Are honeybees suitable surrogates for

use in pesticide risk assessment for non-apis bees? Pest Management Science. 2019;**75**(10):2549–2557. DOI: 10.1002/ps.5494

[155] Sentil A, Lhomme P, Reverté S, et al. The pollinator conservation approach "farming with alternative pollinators": Success and drivers. Agriculture, Ecosystems & Environment. 2024;**369**:109029. 109029. DOI: 10.1016/j.agee.2024.109029

[156] Boyle NK, Pitts-Singer TL, Abbott J, et al. Workshop on pesticide exposure assessment paradigm for non-apis bees: Foundation and summaries. Environmental Entomology. 2018;**48**(1):4–11. DOI: 10.1093/ee/nvy103

[157] Axelman J, Aldrich A, Duquesne S, et al. A systems-based analysis to rethink the European environmental risk assessment of regulated chemicals using pesticides as a pilot case. Science of the Total Environment. 2024;**948**:174526. 174526. DOI: 10.1021/acs.est.5c08032

[158] Rosa-Fontana A, Aguado-López D, Uriel CJ, et al. Bee gut microbiota as an emerging endpoint for the environmental risk assessment of pesticides. Science of the Total Environment. 2025;**993**:179977. 179977. DOI: 10.1016/j.scitotenv.2025.179977

[159] Vryzas Z. S16-02 new approach methods in risk assessment of biopesticides. Toxicology Letters. 2025;411. DOI: 10.1016/j.toxlet.2025.07.095

[160] Nehul JN. Environmental impact of pesticides: Toxicity, bioaccumulation and alternatives. Environmental Reports. 2025;7(2):14–21. DOI: 10.51470/er.2025.7.2.14

[161] Brewer L, Warren-Hicks W, Hinarejos S, et al. A global nectar and pollen pesticide residue database with a user interface tool for calculating residue per unit dose for different pesticide application methods. Integrated Environmental Assessment and Management. 2025;**21**:1471–1481

[162] Liu X, Lu Y, Zhang X, et al. Biosensors for detection of pesticide residues. Trends in Analytical Chemistry. 2020;**122**:115701. DOI: 10.1016/j.trac.2019.115701

[163] Rusman MAA, Muhsin MA, Akbar A. Eco-friendly Pesticide Innovation: A Sustainable Pest Control Strategy for Modern Agriculture. In: International Conference on Agricultural and Biological Engineering 2025. 13 May, 2025. Vol. **180**. Makassar, Indonesia; 01011 p. DOI: 10.1051/bioconf/202518001011

[164] Marrone PG. The market and potential for biopesticides. 246th National Meeting of the American-Chemical-Society (ACS). September 08-12, 2013. **1172**. Indianapolis, IN. 2014;245–258 p

[165] Areej A, Usama M, Zulfiqar U, et al. Biopesticides in sustainable agriculture: Enhancing targeted pest control and ecosystem health. Applied Agricultural Science. 2024;**2**(1):1–8. DOI: 10.25163/agriculture.2110006

[166] Pecenka JR, Ingwell LL, Krupke CH, et al. Implementing IPM in crop management simultaneously improves the health of managed bees and enhances the diversity of wild pollinator communities. Scientific

Reports. 2023;**13**(1). DOI: 10.1038/s41598-023-38053-5

[167] Tonle FBN, Niassy S, Ndadji MMZ, et al. A road map for developing novel decision support system (DSS) for disseminating integrated pest management (IPM) technologies. Computers and Electronics in Agriculture. 2024;**217**:108526. 108526. DOI: 10.1016/j.compag.2023.108526

[168] Zhou W, Arcot Y, Medina RF, et al. Integrated pest management: An update on the sustainability approach to crop protection. ACS Omega. 2024;**9**(40):41130–41147. DOI: 10.1021/acsomega.4c06628

[169] Phan NT, Biddinger DJ, Rajotte EG, et al. Pesticide use in integrated pest and pollinator management framework to protect pollinator health. Pest Management Science. 2024;**81**(4):1691–1696. DOI: 10.1002/ps.8582

[170] Phan NT, Rajotte EG, Smagghe G, et al. Agricultural pesticide regulatory environment for pollinator protection across geographical regions. Frontiers in Sustainable Food Systems. 2023:7. DOI: 10.3389/fsufs.2023.1241601

[171] Karbassioon A, Stanley DA. Exploring relationships between time of day and pollinator activity in the context of pesticide use. Basic and Applied Ecology. 2023;**72**:74–81. DOI: 10.1016/j.baae.2023.06.001

[172] Gurr GM, Wratten SD, Luna JM. Multi-function agricultural biodiversity: Pest management and other benefits. Basic and Applied Ecology. 2003;**4**(2):107–116. DOI: 10.1078/1439-1791-00122

[173] EFSA. Evaluation of the data on clothianidin, imidacloprid and thiamethoxam for the updated risk assessment to bees for seed treatments and granules in the EU. EFSA Supporting Publications. 2018;**15**(2):1378E–1378E

[174] Schatz B, Maxime D, Mickael H, et al. Pollinator conservation in the context of global changes with a focus on France and Belgium. Acta Oecologica. 2021;**112**:103765. 103765. DOI: 10.1016/j.actao.2021.103765

[175] Nalepa R, Colla SR. Toward a wild pollinator strategy for Canada: Expert-recommended solutions and policy levers. Facets. 2023;**8**:1–18. DOI: 10.1139/facets-2022-0204

[176] Whitney KS, Stringer BB. Evaluation of us state pollinator plans using 3 evidence-based policymaking frameworks. bioRxiv. 2021. DOI: 10.1101/2021.06.15.447774

[177] Durant JL, Ponisio LC. A regional, honey bee-centered approach is needed to incentivize grower adoption of bee-friendly practices in the almond industry. Frontiers in Sustainable Food Systems. 2021;**5**. DOI: 10.3389/fsufs.2021.628802

[178] Dicks LV, Viana B, Bommarco R, et al. Ten policies for pollinators. Science. 2016;**354**(6315):975–976. DOI: 10.1126/science.aai9226

[179] Cole LJ, Kleijn D, Dicks LV, et al. A critical analysis of the potential for EU common agricultural policy measures to support wild pollinators on farmland. Journal of Applied Ecology. 2020;**57**(4):681–694. DOI: 10.1111/1365-2664.13572

[180] Rudnicki R, Wiśniewski Ł, Biczkowski M. A spatial typography of environmentally friendly common agricultural policy support relevant to European green deal objectives. Land. 2021;**10**(10):1092. DOI: 10.3390/land10101092

[181] Montana J. How IPBES works: The functions, structures and processes of the intergovernmental platform on biodiversity and ecosystem services. SSRN Electronic Journal. 2016. DOI: 10.2139/ssrn.2778701

[182] Mukhtar Y, Shankar UMA. Integrated pest and pollinator management in India: A way forward to sustainable agriculture. The Indian Journal of Agricultural Sciences. 2023;**93**(9):939–947. DOI: 10.56093/ijas.v93i9.134403

[183] Sharma S. Cultivating sustainable solutions: Integrated pest management (IPM) for safer and greener agronomy. Corporate Sustainable Management Journal. 2023;**1**(2):103–108. DOI: 10.26480/csmj.02.2023.103.108

[184] Constantine KL, Kansiime MK, Mugambi I, et al. Why don't smallholder farmers in Kenya use more biopesticides? Pest Management Science. 2020;**76**(11):3615–3625. DOI: 10.1002/ps.5896

[185] Stanley DA, Raine NE, Ayasse M. Chronic exposure to a neonicotinoid pesticide alters the interactions between bumblebees and wild plants. Functional Ecology. 2016;**30**(7):1132–1139. DOI: 10.1111/1365-2435.12644

[186] Christen V, Mittner F, Fent K. Molecular Effects of Neonicotinoids in Honey Bees (Apis mellifera). Environmental Science & Technology. 2016;**50**(7):4071–4081. DOI: 10.1021/acs.est.6b00678

[187] Li B, Ke L, Li AR, et al. Exposure of larvae to sublethal thiacloprid delays bee development and affects transcriptional responses of newly emerged honey bees. Frontiers in Insect Science. 2022:2. DOI: 10.3389/finsc.2022.844957

[188] Vandereyken K, Sifrim A, Thienpont B, et al. Methods and applications for single-cell and spatial multi-omics. Nature Reviews Genetics. 2023;**24**(8):494–515. DOI: 10.1038/s41576-023-00580-2

[189] Grozinger CM, Zayed A. Improving bee health through genomics. Nature Reviews Genetics. 2020;**21**(5):277–291. DOI: 10.1038/s41576-020-0216-1

[190] Main AR, Hladik ML, Webb EB, et al. Beyond neonicotinoids – Wild pollinators are exposed to a range of pesticides while foraging in agroecosystems. Science of the Total Environment. 2020;**742**:140436. 140436. DOI: 10.1016/j.scitotenv.2020.140436

[191] Tan K, Radcliffe CM, Wang Z. Gene editing for bee health: Potential applications and ethical considerations. Current Opinion in Insect Science. 2022;**50**:100895. DOI: 10.1016/j.cois.2022.100895

[192] Brittain CA, Vighi M, Bommarco R, et al. Impacts of a pesticide on pollinator species richness at different spatial scales. Basic and Applied Ecology. 2010;**11**(2):106–115. DOI: 10.1016/j.baae.2009.11.007

[193] Vanbergen AJ, Baude M, Biesmeijer JC, et al. Threats to an ecosystem service: Pressures on pollinators. Frontiers in Ecology and the Environment. 2013;**11**(5):251–259. DOI: 10.1890/120126

[194] Thompson HM. Behavioural effects of pesticides in bees - their potential for use in risk assessment. Ecotoxicology. 2003;**12**(1-4):317–330. DOI: 10.1023/A:1022575315413

[195] European Food Safety. Authority, Conclusion on the peer review of the pesticide risk assessment for bees for the active substance clothianidin. EFSA Journal. 2013;**11**(1). DOI: 10.2903/j.efsa.2013.3066

[196] USEPA (United States Environmental Protection Agency). BeeREX v1.0: A screening tool for estimating risks to honey bees from pesticides. [Internet] 2020; Available from: https://www.epa.gov/pollinator-protection/beerex [Accessed: 2025-August-15]

Chapter 4

Anthropogenic Impact to Stream and Groundwater Quality on the Island of Oahu, Hawaii

Steven R. Spengler and Samuel C. Spengler

Abstract

This chapter provides an overview of stream and groundwater quality on the island of Oahu, Hawaii, spanning the past fifty years. Water samples were collected from wells and streams in watersheds characterized by agricultural, residential and mixed land use. Streams were sampled under both baseflow (groundwater-dominant) and runoff conditions. The average concentrations of total nitrogen, phosphorus, wastewater components (caffeine and carbamazepine), and the pesticide glyphosate measured in streams increase from five- to ten-fold under runoff conditions compared to baseflow levels. High-frequency stream sampling revealed that over 90% of the total annual loads of nutrient, glyphosate, and wastewater constituent loads entering near-shore waters are transported during large episodic rainfall events. The concentration levels of soil fumigants (trichloropropane and 1,2-dibromo-3-chloropropane) and the organochlorine pesticide dieldrin, chemicals that have been banned since the mid-1980s, have continued to rise in both groundwater and surface water over the last fifty years. Groundwater and surface water in former agricultural areas tend to have elevated levels of nitrate and certain trace metals (vanadium and silica). Of particular concern, higher vandium levels in groundwater have been statistically linked to increased occurences of nontuberculous mycobacterial (NTM) pulmonary disease among Central Oahu residents. This association suggests a potential environmental health risk, highlighting the need for ongoing monitoring, public health awareness, and possible mitigation efforts to protect community well being.

Keywords: Oahu, Hawaii, stream contamination, groundwater contamination, baseflow, runoff, temporal trends in contamination, anthropogenic impact, pesticides, glyphosate, dieldrin, fumigants, nutrients, solvents, vanadium, wastewater, nontuberculous mycobacterial pulmonary disease, fumigants

1. Introduction

The quality of groundwater and surface waters on the island of Oahu has been negatively affected by urban and commercial agricultural activities over the past 100 years. The large-scale, year-round application of nutrients, herbicides, and termiticides on agricultural and residential lands in close proximity to streams and

IntechOpen

major groundwater aquifers on Oahu has directly affected water quality. For example, fish captured in urban Oahu streams contained the highest concentration levels of pesticides measured in a nationwide water quality study conducted by the United States Geological Survey (USGS) in the 1990s [1].

Large-scale plantation agriculture expanded quickly on Oahu at the start of the 20th century, driven by a population boom as immigrants arrived to work on the burgeoning sugar plantations. The island poulation doulbled from 40,000 to 82,000 between 1896 and 1910. The footprint of commercial agriculture on Oahu peaked before and after World War II, with around 43,000 acres in sugarcane cultivation and as much as 28,000 acres in pineapple production. During this era, Hawaii produced about 75% of the world's pineapple crop and 25% of the United States' domestically grown sugar crop [2]. By 1950, the population on Oahu had grown to 353,000 residents, driven by the growth of both agriculture and military infrastructure on the island post-World War II. However, by the end of the twentieth century, large-scale plantation agriculture on Oahu had largely ended, replaced by expanded urban development and significant growth in tourism.

Today, two pineapple plantations continue to operate on 3,400 acres of land in Central Oahu, and the island's population has grown to over one million residents [3]. Seed crops have become Hawaii's most valuable agricultural product, with several companies acquiring previously abandoned agricultural lands to develop genetically modified cultivars engineered for higher productivity and pest resistance.

The impact of legacy plantation agriculture and the expansion of urban and military development on Oahu over the past 70 years is reflected in the persistence of pesticide, nutrient, and metals contamination in the island's streams and groundwater. These contaminants originated from both point and nonpoint sources. Nonpoint sources include the historical application of pesticides and fertilizers to agricultural fields and the ongoing use of pesticides around homes and lawns to protect structures from termites and to manage weeds. Point sources, on the other hand, consist of industrial sites, underground storage tanks, motor pools, cesspools and landfills.

Total precipitation on the island of Oahu is estimated to be 1,836 million gallons per day (MGD) based on meteorological data collected between 1978 and 2007 [4]. Recharge to Oahu aquifers is estimated to be 597 MGD, while island-wide runoff via streams is 264 MGD, and total evapotranspiration is estimated to be 1,066 MGD [4].

On Oahu, there are only 29 perennial streams that reach the ocean. Waikele is the largest stream on Oahu, with a median annual discharge of around 35 cubic feet per second (CFS) entering the ocean, while most other perennial streams on Oahu have median discharges of less than 5 CFS. The flashy nature of runoff and the paucity of large perennial streams precludes large-scale development of surface water on Oahu and have led to near-total reliance on groundwater for water supply. Oahu's municipal water system produces about 145 MGD of water from 194 groundwater wells, five shafts, and 13 water tunnels to serve the island's roughly 1.1 million residents and visitors.

Surface flow in the small perennial streams on Oahu returns to shallow groundwater-dominated input within a day or two of moderate rainfall events [5], while larger streams such as Waikele return to groundwater-dominated conditions within a week after the cessation of large rainfall events. Contaminant concentrations measured in streams under baseflow conditions reflect the composite concentration of a contaminant present in the top of the aquifer underlying the

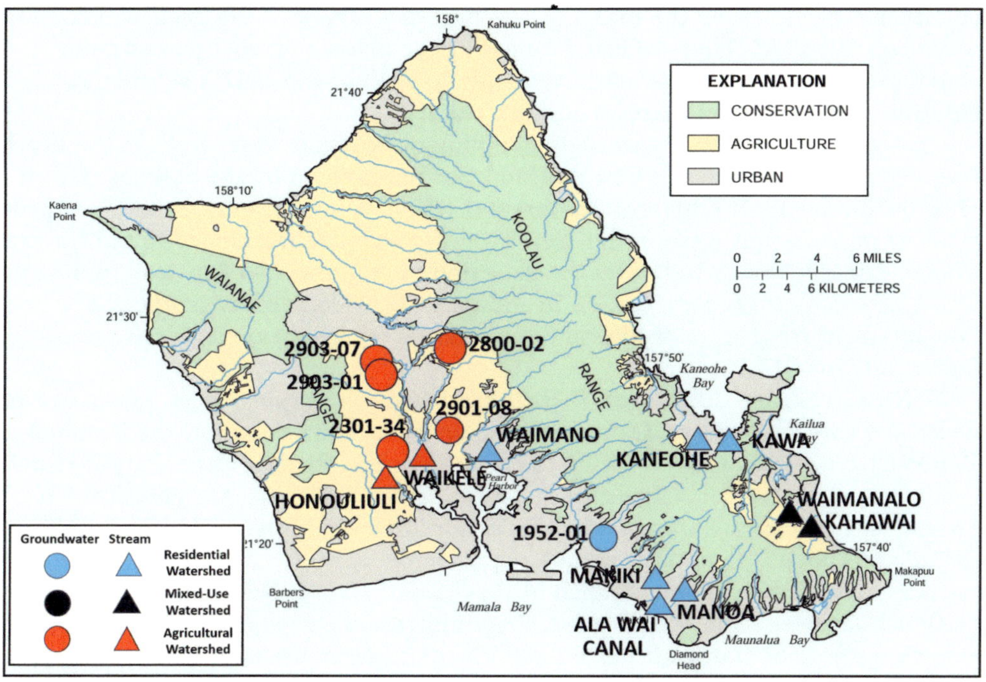

Figure 1.
Land Use of Groundwater and Stream Monitoring Locations.

watershed contributing to stream baseflow. In contrast, contaminant concentrations measured in groundwater collected from water supply and monitoring wells on Oahu reflect a contaminant's concentration present in the immediate vicinity of the sampled well.

The amount and nature of contaminants present in groundwater and streams are expected to vary as a function of the land-use type within the watershed (in the case of a stream) or in the surrounding intermediate area (in the case of a well). On Oahu, stream and groundwater data can be characterized into three general land-use categories: residential, agricultural, and mixed use (**Figure 1**). Stream contaminant data are also characterized into data collected under baseflow conditions and data collected during storms.

2. Previous studies

Researchers from the Department of Agricultural Biochemistry at the University of Hawaii were the first to measure pesticide levels in streams and groundwater on Oahu between 1970 and 1971 [2]. Potable water was collected from 45 drinking water sources (wells, shafts, tunnels, and springs), and nonpotable water was collected from two canals, one urban stream, and the sewage outfall at Sand Island. Dieldrin, chlordane, lindane, p,p'-DDD, p,p'-DDT, and pentachlorophenol (PCP) were detected in the low- to mid-parts-per-trillion (PPT) range.

The Water Resources Research Center (WRRC) at the University of Hawaii conducted sampling of streams, bay waters, and stream and bay sediments for chlorinated pesticides in the mid-1970s as part of a survey of the quality of coastal waters on Oahu [6]. They collected samples from urban, mixed-use, and rural watersheds and detected low parts-per-trillion levels of p,p'-DDT, chlordane, dieldrin, and PCP in both stream and bay waters.

During the early 1980s, extensive sampling of drinking water wells in the Pearl Harbor groundwater area in Central Oahu was undertaken by the State of Hawaii after the discovery of ethylene dibromide (EDB) and 1,2-dibromo-3-chloropropane (DBCP) in numerous wells in the area. This discovery ultimately led to the closure of nine drinking-water wells and the detection of an impurity of the soil fumigant DD, 1,2,3-trichloropropane (TCP), in additional wells located hydraulically downgradient from areas of pineapple cultivation where these chemicals were historically used [7].

Between 1999 and 2001, the USGS sampled 30 public-supply wells, 15 monitoring wells, and three streams on Oahu between 1999 and 2001 as part of the National Water-Quality Assessment (NWQA) program [1]. The streams were analyzed for 47 pesticides, as well as for suspended sediment, major ions, and nutrients. The groundwater samples were analyzed for major ions, nutrients, pesticides, volatile organic compounds, and trace metals.

The Safe Drinking Water Branch of the State of Hawaii Department of Health (HDOH) has been compiling organic contaminant data measured in drinking water supply wells statewide since 1993. The database consists of data collected by both the Department of Health and other agencies, which have been confirmed through repeat testing. The compiled contaminant data can be viewed on maps and downloaded from the following website: https://eha-cloud.doh.hawaii.gov/sdwb/#!/home.

The HDOH collected surface water and sediment samples from 24 sites around the state between December 2013 and January 2014, which were analyzed by the USGS [8]. The samples were analyzed for a total of 136 pesticides or breakdown products in surface water and 121 pesticides in sediment.

Between 2015 and 2017, the Surfrider Foundation conducted a multi-year water quality study evaluating the concentration of pesticides and herbicides in the nearshore marine environments on the islands of Oahu and Kauai [9]. A total of 59 water samples and 32 colocated sediment samples were collected between 2015 and 2017 and samples were analyzed for a total of 197 pesticide compounds, including glyphosate.

The USGS and Hawaii Department of Agriculture conducted a comprehensive pesticide-monitoring program sampling surface water across the state of Hawaii between 2016 and 2019 [10]. This study generated data on the concentration levels of pesticides in 21 streams that flow through agricultural, residential, and mixed-use areas on the island of Oahu under baseflow and storm runoff conditions. The surface-water samples were analyzed for 253 pesticide compounds, but not for the most ubiquitous pesticide in the state, glyphosate.

Extensive sampling for glyphosate in Oahu streams across residential, mixed-use, and agricultural areas has been conducted under both baseflow and runoff conditions over the past decade in separate studies [11, 12]. A total of 129 baseflow and 231 runoff samples have been collected over this 10-year period, along with 81 streambed sediment samples and nine suspended sediment samples. All samples were analyzed for glyphosate, and a subset of the sediment samples collected was analyzed for its degradation product, aminomethylphosphonic acid, to evaluate the distribution and persistence of glyphosate in the Hawaiian environment.

The Safe Drinking Water Branch collected samples of wastewater influent entering and effluent exiting five wastewater treatment plants on Oahu in 2017 [13]. These samples were analyzed for the presence of both pesticides and pharmaceutical compounds.

A 2022 study measured the spatial distribution of two wastewater components (carbamazepine and caffeine) present in the dense network of storm water manholes located in the coastal tourist destination of Waikiki, in the adjacent inland McCully-Moiliili residential areas, and in the industrial area of Mapunapuna between November 2020 and December 2021 [14]. These data were used to identify locations of ongoing sewage exfiltration in the Waikiki area. In another study, 36 streams and 11 springs on Oahu were repeatedly sampled for nutrients and wastewater components (carbamazepine, sulfamethoxazole, and caffeine) under low baseflow conditions between November 2020 and September 2022 [5]. Additional sampling was conducted for these pharmaceutical compounds between January 2024 and April 2025 under runoff conditions. Sampling from public septic tanks was conducted between 2020 and 2025 to determine the average concentration levels of caffeine, carbamazepine, and sulfamethoxazole present in wastewater.

Researchers at the USGS conducted vertical profile sampling of two wells in the island's largest aquifer, the Pearl Harbor aquifer, to evaluate variations in geochemical and isotopic properties with depth [15]. The observed geochemical variations indicate that the upper 75 to 125 meters of the basal aquifer consists of water recharged from local rainfall and irrigation over the past few decades. Below this upper anthropogenically impacted layer is the core of the freshwater lens, which is between 100 to 150 meters thick and has an apparent carbon-14 age of 1,800 years. This freshwater lens floats on a saltwater layer with an apparent carbon-14 age of between 6,000 and 9,000 years.

The impact on the water quality of the upper portions of the Pearl Harbor aquifer basal lens is illustrated by the dramatic difference in synoptic water quality measured at two USGS monitoring stations located on Waikele Stream (USGS-16213000 and USGS-212402158010501) 1.4 miles apart (**Table 1**). Under baseflow conditions, the stream at the upper monitoring station is predominantly composed of high-level water originating from dike-impounded spring flow at higher elevations in the Koolau Mountains. The section of the stream between the two monitoring locations is gaining due to the bottom of Waikele Stream intercepting the top of the basal Pearl Harbor aquifer. As a result, the stream water reaching the lower monitoring station is composed of roughly 88% spring discharge from the top of the basal groundwater lens. The large increases in nutrient and metal levels measured between these two sections of the same stream illustrate the significant anthropogenic impact of agricultural use of Central Oahu lands on the water quality present in the upper portion of the basal groundwater body underlying Central Oahu (**Table 1**).

3. Nutrients

The daily load of nutrients exiting the major watersheds on Oahu into the ocean was estimated using the Pollutant Loading Application (PLOAD) model developed by the U.S. Environmental Protection Agency [16]. The average daily load of

Analyte/ measurement	USGS – 212402158010501: 1.4 miles upstream	USGS – 16213000 station	Calculated spring input composition[1]	Units	Present enrichment in agricultural return impacts spring water over high-level groundwater input
Streamflow	2.12	18	NA	cfs	
Nitrate	0.037	1.62	1.83	mg/L	4,951%
Vanadium	1.5	28	32	ug/L	2,103%
Total nitrogen	0.11	1.7	1.91	mg/L	1,739%
Sulfate	2.2	17.4	19	mg/L	883%
Chloride	11	86	96	mg/L	873%
Phosphorous	0.027	0.184	0.21	mg/L	759%
Sodium	8	54	60	mg/L	752%
Specific conductance	111	487	537	uS/cm	484%
Boron	21	90	99	ug/L	473%
Magnesium	3	11	12	mg/L	402%
Barium	1.7	6.1	6.7	ug/L	393%
Silica	14	49	54	mg/L	383%
Strontium	32	91	99	ug/L	309%
Potassium	860	2,300	2,500	ug/L	291%
Kjeldahl nitrogen	70	40	36	ug/L	51%

[1]*Calculated assuming that water at downstream station 1,213,000 is composed of 88.2% basal spring flow and 11.8% high-level groundwater input measured at the upstream site.*

Table 1.
Analyte concentration in Waikele Stream at two locations showing enrichment in irrigation-return-impacted groundwater.

total suspended solids, total nitrogen, nitrate, ammonia, and total phosphorus exiting the 87 watersheds via streams and/or runoff on Oahu was estimated to be 178,600 kg/day, 2,305 kg/day, 229 kg/day, 85.5 kg/day, and 1,090 kg/day, respectively.

Extensive high-frequency sampling has been conducted in the 4.05-square-kilometer residential Kawa watershed in windward Oahu since 2017. **Table 2** shows the total mass of suspended sediment, nutrients, and glyphosate measured leaving the Kawa watershed and entering Kaneohe Bay during eight runoff events, where a minimum of eight stream samples were analyzed for nutrients and glyphosate during each individual runoff event. The TSS load was calculated from continuous turbidity values measured at the USGS monitoring station on Kawa Stream. The average ratio of the mass of TSS discharging from the Kawa watershed relative to the mass of nitrogen and phosphorus measured during these eight runoff events is higher than the island-wide ratio calculated by the PLOAD model.

Figures 2 and **3** present the concentration levels of total nitrogen and total phosphorus measured in six streams across Oahu. Kawa, Manoa, Makiki, and Kaneohe streams represent water quality produced from residential watersheds. Waimanalo Stream represents water quality produced from a mixed-use watershed,

Rainfall event start date	Runoff event duration (hours)	Max Streamflow (cfs)	TSS (kg)	Total nitrogen (kg)	Nitrate (kg)	Ammonia (kg)	Phosphorous (kg)	Glyphosate (kg)
2/5/2018	13.0	745	35,760	192	20	8.0	90	0.011
4/26/2019	3.8	572	13,682	78	13	5.0	41	0.092
4/27/2019	3.0	162	514	10	2.7	0.9	2.2	0.006
4/29/2019	3.0	286	5,464	28	14.0	1.0	8.0	0.019
5/3/2019	2.5	417	3,555	26	5.0	1.1	13.4	0.020
8/7/2019	3.0	290	4,590	18	5.2	1.2	4.6	0.011
3/15/2020	9.0	1,220	94,702	568	47	5.0	166	0.035
5/13/2024	11.5	1,600	314,805	1,123	176	35	636	0.066

Table 2.
TSS, nutrient, and glyphosate mass during storm events.

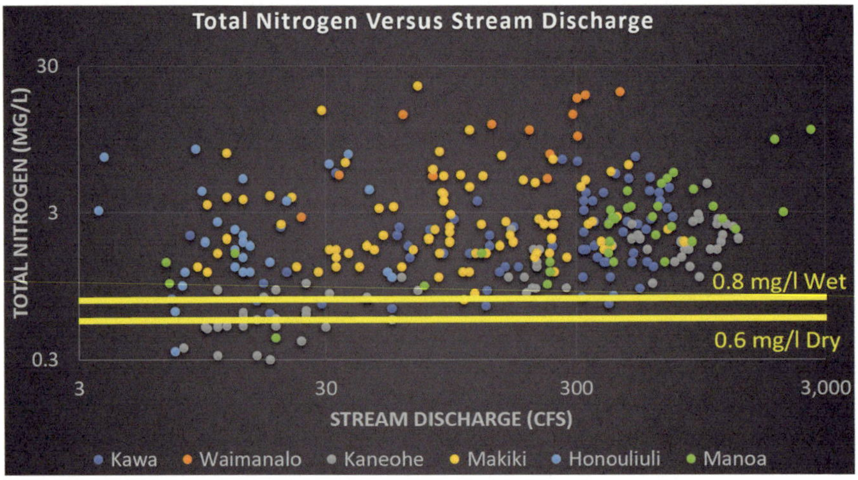

Figure 2.
Total nitrogen concentration versus stream discharge.

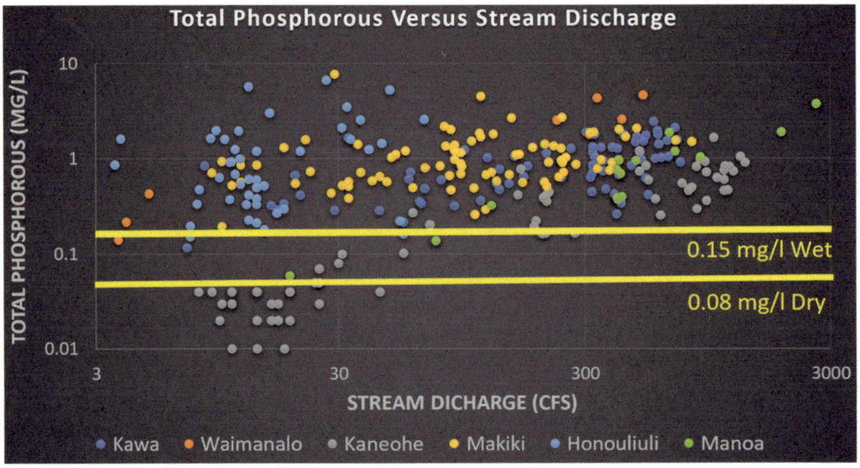

Figure 3.
Total phosphorus concentration versus stream discharge.

while Honouliuli Stream flows through an agricultural watershed. Honouliuli Stream generally had the highest nitrogen and phosphorus levels at low flow rates, while Waimanalo Stream had the highest overall total nitrogen levels under runoff conditions. The concentration data are plotted against the 2% not-to-exceed water quality standard (WQS) regulatory levels developed for the State of Hawaii for both dry and wet weather conditions. Total nitrogen and total phosphorus concentrations in most Oahu streams are significantly higher than the existing 2% Hawaii WQS under both baseflow and high-flow runoff conditions. In contrast to total nitrogen and phosphorous, stream nitrate levels generally decline at higher flows. Hawaii does not have State WQs for nitrate. The highest levels of nitrates on Oahu are measured in wells and springs located hydraulically downgradient of lands formerly used to grow sugarcane in the Pearl Harbor and Waialua aquifers of Central Oahu [5].

Current implementation of the WQS results in over 90% of streams and ocean monitoring sites on Oahu reporting at least one exceedance of nutrient and/or microbial standards in biennial State of Hawaii Water Quality Monitoring and Assessment reports. [17]. For instance, the most recent biennial assessment found that 93% of the marine water bodies assessed (137 out of 148) exceeded WQSs for one or more parameters. This suggests that the current WQS for Hawaiian streams and marine water bodies should be updated to reflect natural variations in nutrient, microbiological, and physical property (TSS and turbidity) concentration levels under baseflow and runoff conditions. Revising the WQS to reflect actual watershed and stream chemistry would enable more precise identification of pollution sources and improve the targeting of mitigation strategies [5].

4. Pesticides

Pesticides, including both insecticides and herbicides, have been applied to agricultural lands on the island of Oahu since the early 1900s. The annual amounts of pesticides historically applied to sugarcane and pineapple crops in Hawaii may have been as much as five times the amounts applied to major temperate region field crops elsewhere in the United States [6]. The intensive use of pesticides, year-round cultivation practices, and the proximity of agricultural lands to riparian areas increased the likelihood of pesticide transport to Oahu streams [18]. As a result, fish caught from urban Honolulu streams in the 1990s had the highest levels of pesticides measured in the United States [1].

The concentration of pesticides present in Oahu groundwater and streams was first quantified in 1971 [2]. At the time, the average detected concentration of chlordane, dieldrin, p,p'-DDT, and lindane in 45 potable water samples from Oahu was around 1 ng/L. The average concentration of dieldrin, p,p'-DDT, and PCP detected in Kalihi Stream was 9.9, 3.3, and 655 ng/L, respectively. PCP was a commonly used herbicide in agricultural fields up until the late 1960s and was also extensively used to render lumber in Hawaii termite-resistant. The concentration levels of dieldrin, p,p'-DDT, and PCP measured in a 24-hour composite sample collected from the Sand Island Wastewater Treatment Plant in 1971 were 198, 107, and 2,600 ng/L, respectively. The average concentration of chlordane, dieldrin, and p,p'-DDT measured in the brackish water in the Ala Wai Canal was 9.1, 11.1, and 1.8 ng/L, respectively. In 1973, researchers at the WRRC at the University of Hawaii measured the average concentration of chlordane, dieldrin, p,p'-DDT, and PCP in Kaneohe and Maunalua Bay waters to be 6.2/0.4, 6.0/0.6, 7.2/1.4, and 36.6/7.6 ng/L, respectively [6].

Sediment collected from three areas in the Ala Wai Canal, which receives sediments from the residential communities of Manoa and Palolo, contained average chlordane and dieldrin levels of 378 and 43 μg/kg in 1971 [2]. Somewhat lower average chlordane and dieldrin concentrations (86 and 19 μg/kg, respectively) were measured in sediments collected from five areas in the Ala Wai Canal roughly 40 years later, in 2010 [19]. In comparison, the average chlordane, dieldrin, DDT, and PCP concentrations measured in sediments collected from Kaneohe and Maunalua Bay in 1972 were significantly lower at 20.5/0.8, 2.4/0.7, 3.5/0.9, and 6.6/ND μg/kg, respectively [6]. The pesticide levels measured in sediment samples in the Ala Wai

Canal, Kaneohe Bay, and Maunalua Bay (Hawaii Kai) were generally 3 to 4 orders of magnitude higher than the levels measured in water samples at the time [6].

The detection of the organochlorine pesticide heptachlor in March 1982 in milk produced on Oahu led to the removal of 36 million pounds of milk from the market. Heptachlor had been applied to pineapples to control mealybug damage. Harvested pineapple plants were used as a cheap local substitute for imported dairy forage. By 1981, the base of the harvested plant material was found to contain high levels of heptachlor. When fed to dairy cows, the pesticide was excreted in the milk. The Hawaiian dairy market never fully recovered from this incident, and today Hawaii imports over 90% of its milk supply.

In 1977, the accidental release of 500 gallons of the soil fumigant EDB within 100 feet of a private water supply well (Kunia Well 2,703–01) led to the testing of water supply wells in Central Oahu for fumigants. It had been previously considered unlikely that the major aquifers in Central Oahu would be impacted by the use of fumigants due to the volatile nature of these chemicals and the presumed protection conferred by the great depth of soil, saprolite, and rock between the land surface and the underlying basal aquifer. However, high concentration levels of EDB (92 to 300 μg/L) and another fumigant DBCP (0.5 to 11 μg/L) were subsequently measured in Well 2,703–01 in 1980 and in other Central Oahu wells located hydraulically downgradient of lands used for pineapple cultivation [7]. Aside from the spill, a subsequent investigation found that spillage during past mixing and transfer operations at chemical storage areas in Central Oahu likely contributed to the observed contamination [20]. This detection of fumigants led to the closure of nine potable water wells in Central Oahu (three in Mililani, four in Waipahu, and two in Kunia) that had been producing 13 mgd of potable water [21]. The detected DBCP and EDB concentration levels were less than 0.1 and 0.2 μg/L at the time the wells were closed. TCP, an impurity in the soil fumigant DD, was also detected in wells downgradient of areas used for pineapple cultivation [7]. **Figure 4** shows the highest TCP concentrations detected in wells and a stream (Waikele) in Central Oahu since the mid-1980s.

Numerical simulations conducted in the 1980s predicted a decline in the DBCP levels to below regulatory action levels (20 ng/L) for impacted wells in Central Oahu by around the year 2000 [21, 22]. In addition, early studies on the presence of fumigants in Central Oahu wells indicated that TCP levels in groundwater might have already been decreasing at the time [7]. Unfortunately, these predictions have proven to be incorrect, as concentration levels of TCP and DBCP have continued to rise since 1983 in monitored wells in Central Oahu (**Figures 5, 6**). It is apparent that the degradation rates and adsorption characteristics measured in the laboratory for these chemicals at the time do not reflect the actual degradation and adsorption rates that occur in water percolating through the thick vadose zone in Central Oahu. Specifically, the adsorption of these chemicals during advection through the saprolite and basalt material, which comprise the vadose zone in Central Oahu, is greater, and the rate of in-situ degradation is slower than initially estimated.

Atrazine is an herbicide that was introduced to Hawaii's sugar industry in 1961 and became the primary herbicide used from the 1960s to the 1980s. Atrazine and its degradation products (OIET) were first detected in groundwater in 1983 at levels ranging from 50 to 170 ng/L, based on testing conducted by the Hawaiian Sugar Planters' Association in wells hydraulically downgradient of lands used for sugarcane cultivation. Atrazine was detected in Waikele Stream at an average concentration level of 6 ng/L in 17 samples collected by the USGS under baseflow (groundwater-dominant)

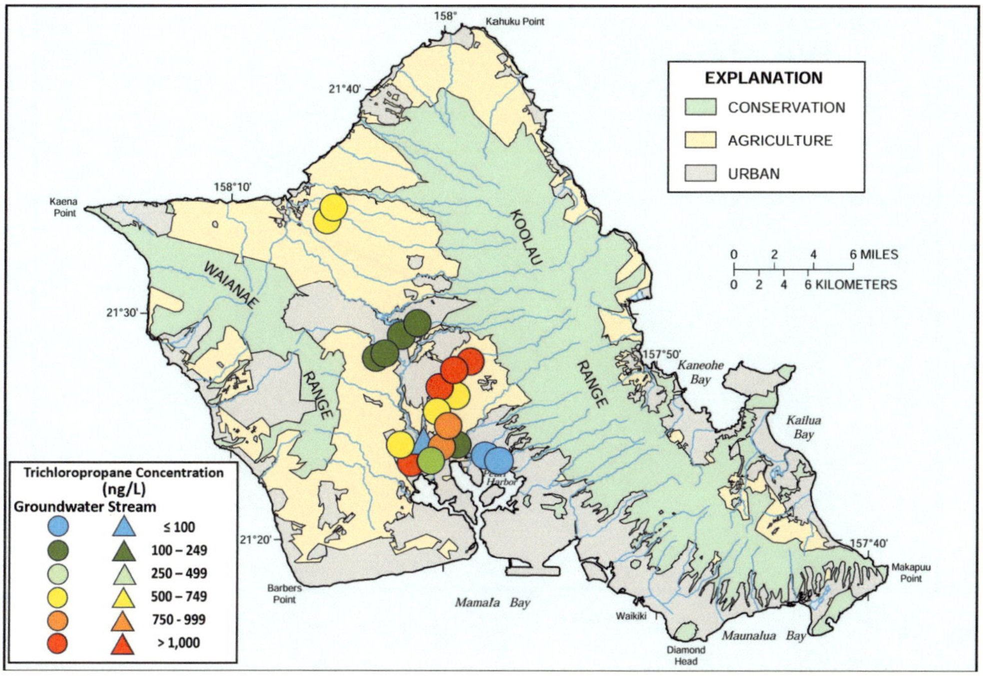

Figure 4.
Location of drinking water wells and stream where TCP was detected.

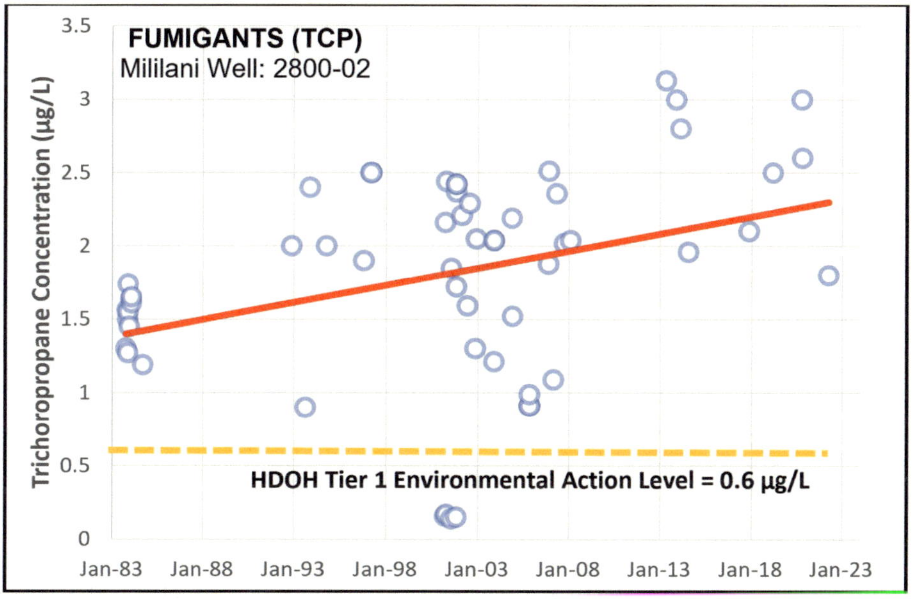

Figure 5.
Trichloropropane concentration in Central Oahu.

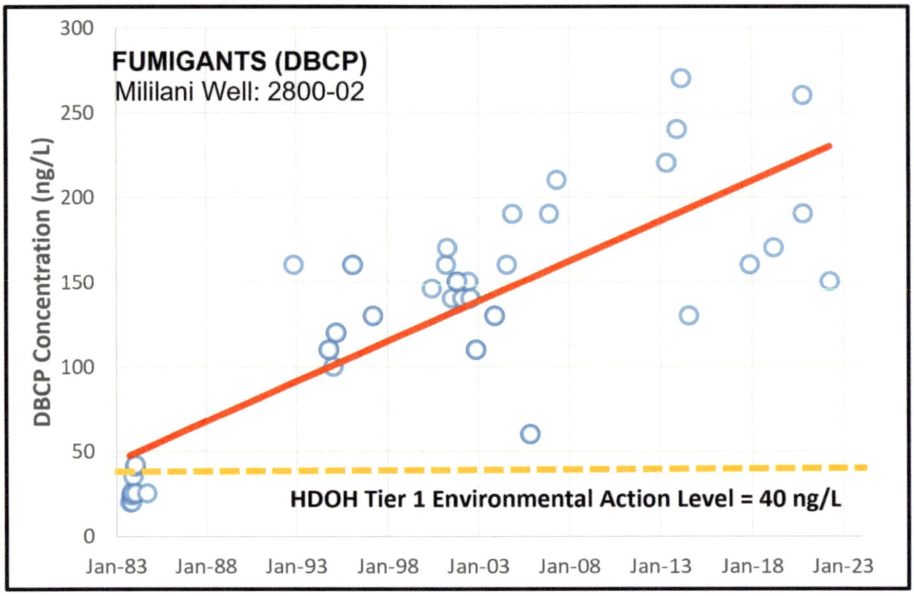

Figure 6.
DBCP concentration in Central Oahu.

conditions between 1999 and 2001 [1]. In addition, 32 other pesticide compounds (22 herbicides, nine insecticides, and one fungicide) were detected in surface-water samples collected from Waikele Stream, which drains agricultural and urban areas in Central Oahu [1]. Unlike the soil fumigants, the concentration levels of atrazine in wells and streams in Central Oahu are currently either nondetectable or declining (**Figure** 7).

Chlordane was the primary organochlorine termiticide used in Hawaii from the mid-1950s to the late 1980s. Chlordane has been detected in five groundwater wells located in urban Honolulu since 1987 at concentration levels between 130 and 530 ng/L. Both chlordane and dieldrin are persistent organic pollutants known to pose risks to human and aquatic ecosystems, which makes their presence in drinking water sources a concern for public safety. Despite being first licensed for use in Hawaii in 1976, dieldrin, another organochlorine insecticide, was also detected in streams, bay waters, and bay sediments in both leeward and windward Oahu during studies conducted in the early 1970s [2, 6]. In comparison to chlordane, dieldrin has been detected in 22 sources (wells and shafts) that provide drinking water to the Pearl Harbor-Honolulu water system. Average detected dieldrin concentrations in these water sources since 1997 have ranged from 15 ng/L to 37 ng/L. **Figure 8** shows the locations of the drinking water supply wells and streams on Oahu where dieldrin has been detected. The use of dieldrin as a termiticide in Hawaii was discontinued in 1989.

The HDOH has established a human health advisory for dieldrin in drinking water of 200 ng/L, which is based on a 1 in 10,000 excess cancer risk for humans. Hawaii's human health advisory for dieldrin in drinking water is significantly higher than health advisories developed for other states, including Minnesota (6 ng/L) and Florida (2 ng/L). The Tier 1 Environmental Action Level (EAL) for dieldrin, developed by HDOH based on drinking water toxicity, is 3.5 ng/L, while the EAL

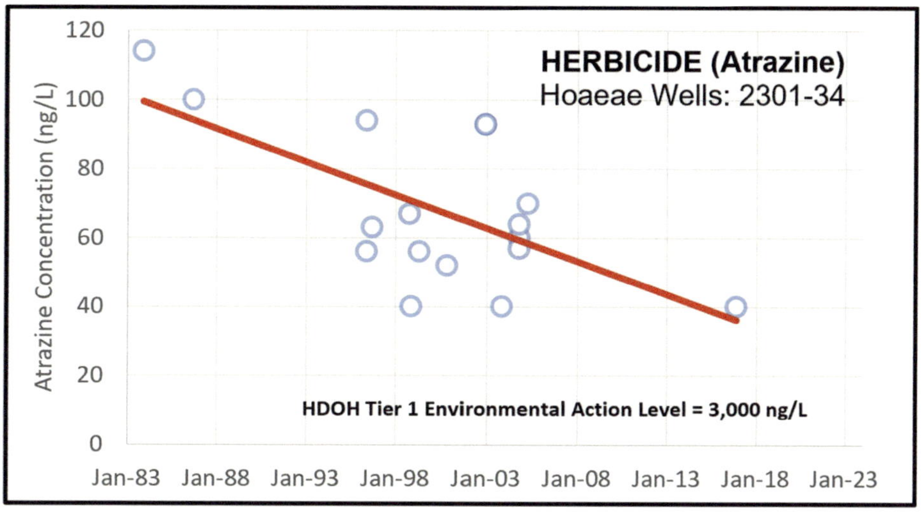

Figure 7.
Atrazine concentration since 1983 in the Pearl Harbor Aquifer.

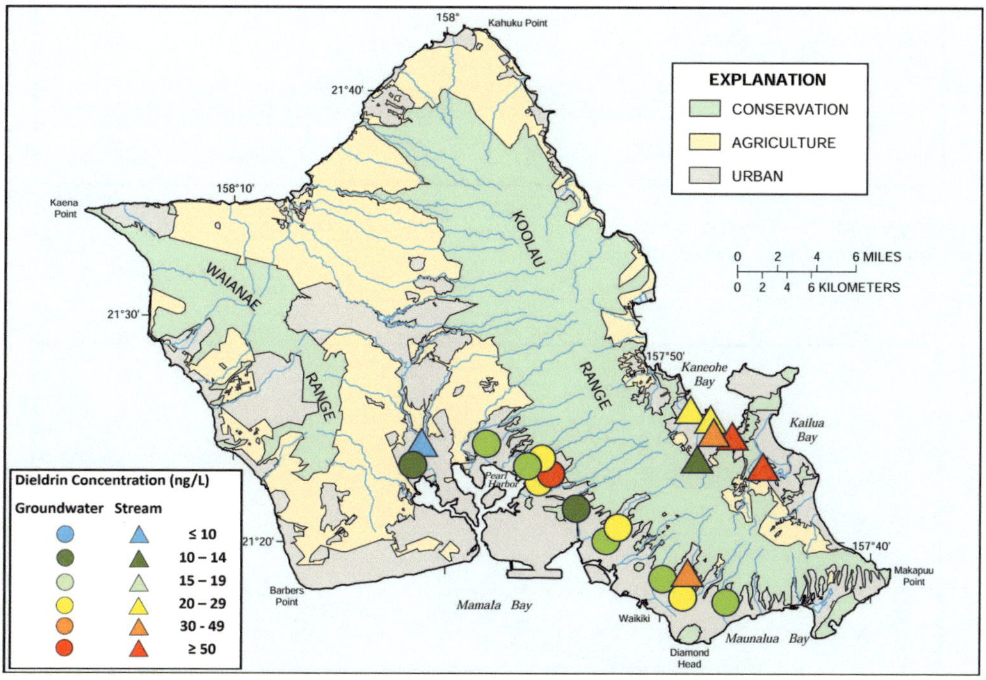

Figure 8.
Drinking water wells and streams where dieldrin has been detected.

based on aquatic criteria is 1.9 ng/L. The ubiquitous occurrence of dieldrin in streams and groundwater on Oahu may pose a risk to organisms in coastal bays such as Pearl Harbor and Kaneohe Bay, which have been receiving a constant input of this chemical since at least the early 1970s. For instance, Pacific oysters in Kaneohe Bay collected from near the mouth of Kaneohe Stream in 1990 contained 200 μg/kg of dieldrin [23].

The concentration level of dieldrin continues to increase in certain water supply wells and streams that drain residential areas on Oahu, despite the use of this insecticide being banned in 1989 in favor of less-persistent compounds. **Figures 9** and **10** show the upward trend in dieldrin concentrations measured over the past 30 years in a groundwater well located in leeward Oahu (**Figure 9**) and over the past 50 years in Kaneohe Stream (**Figure 10**) under baseflow conditions in windward Oahu. The half-

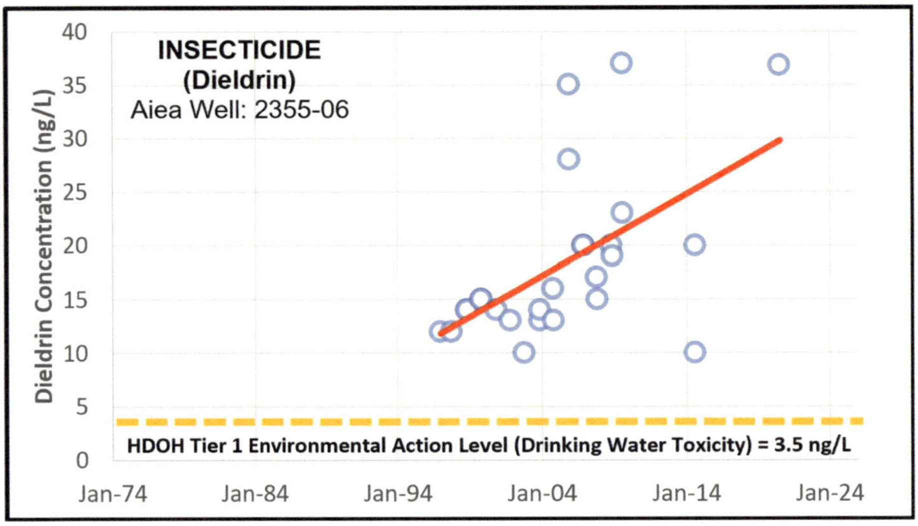

Figure 9.
Dieldrin concentration in an water supply well.

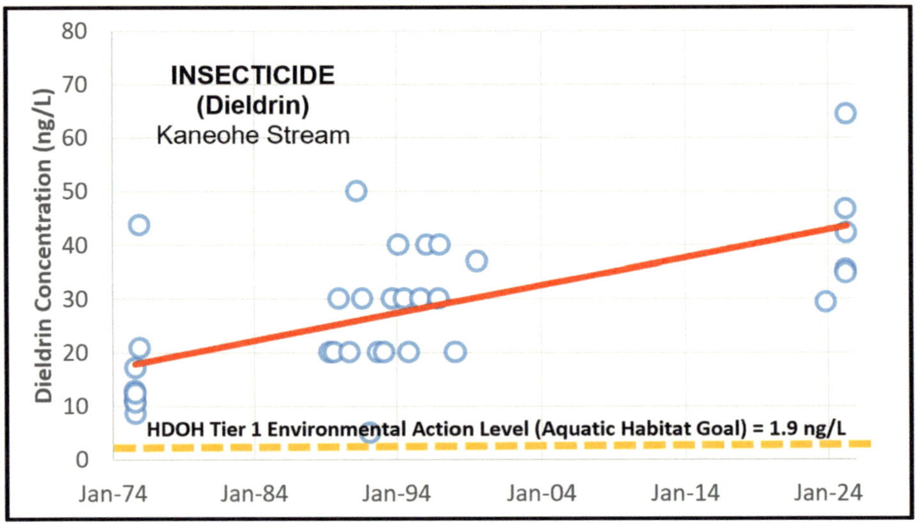

Figure 10.
Dieldrin concentrations in Windward Oahu stream.

life of dieldrin in the environment has been estimated to be around five years [24], but the trends observed in streams and groundwater on Oahu suggest that its actual persistence in tropical environments may last for decades.

The USGS, Hawaii Department of Agriculture, and Surfrider Foundation conducted comprehensive pesticide-monitoring programs of surface water in Hawaii between 2016 and 2019 [9, 10]. These studies generated data on concentration levels of pesticides in 21 streams that flow through agricultural, residential, and mixed-use areas on the island of Oahu under baseflow and runoff conditions. The surface-water samples were analyzed for 253 pesticide compounds in the USGS study, but not for the most ubiquitous pesticide in the state, glyphosate [10]. The Surfrider-funded study analyzed stream samples for 197 pesticide compounds, including glyphosate. Additional high-frequency sampling for glyphosate was conducted between 2015 and 2020 in a number of streams on Oahu [9, 11, 12]. **Table 3** summarizes the average pesticide concentrations measured under both baseflow (groundwater-dominant) and runoff conditions during these studies in seven streams that flow through watersheds dominated by residential, mixed-use, and agricultural land uses on Oahu.

The concentration levels of nonglyphosate herbicides, fungicides, and insecticides detected in Oahu streams are generally in the low nanogram per liter (ng/L) range (< 60 ng/L) and have similar average detected concentrations under both baseflow and runoff conditions. In contrast, glyphosate concentrations were generally detected at mid-ng/L to low microgram per liter (μg/L) levels, with average concentrations measured under runoff conditions being five to 10 times higher than at baseflow. With the exception of dieldrin, pesticide concentrations detected in Oahu streams are generally significantly lower than human-health and acute aquatic-life benchmarks [10].

4.1 Glyphosate

Glyphosate is the active ingredient in the popular weed killer Roundup® and is the most widely used herbicide in the United States. Glyphosate-based herbicides are also the most widely used herbicides globally, with annual consumption volumes projected to reach 740,000–920,000 tons by 2025 [25]. Nonagricultural uses of glyphosate in the United States include turf grass production, residential and commercial areas, parks, and wildlife management areas.

Between 2015 and 2020, the average concentration levels of glyphosate measured in Oahu streams were an order of magnitude higher than the average concentration levels of other commonly detected pesticides measured during the USGS and Surfrider-funded studies [9, 10]. In 2023, Bayer, the sole producer of glyphosate in the United States, stopped using glyphosate in its residential weed-killing products due to ongoing litigation concerns. **Figure 11** shows the generally similar levels of glyphosate measured in Kawa Stream, which flows through a residential neighborhood in windward Oahu, both prior to and after the discontinued use of glyphosate in weed-killing products sold for residential use. The persistence of glyphosate in the Kawa Stream watershed thus far likely reflects the continued residential use of Roundup products by homeowners who purchased them prior to 2024.

Chemical name	Type	Residential watersheds						Mixed-use watersheds				Agricultural watersheds			
		Kawa (ng/L)		Kaneohe (ng/L)		Manoa (ng/L)		Waimanalo (ng/L)		Kahawai (ng/L)		Honouliuli (ng/L)		Waikele (ng/L)	
		Base	Strom	Base	Storm	Base	Storm	Base	Storm	Base	Storm	Base	Storm	Base	Storm
Glyphosate	Herbicides	89	939	25	295	20	1,511	94	456	393	2,652	769	976		
2,4-D		49	208	1,390	92	37							56		51
Triclopyr		51	61	82			264	3			90				
Imazaquin			165	10		11					17				12
Diuron		42	88	45	25	10	26			3	9			12	19
Bromacil		22		13						7				19	77
Simazine						4							1,454		
OIET										4			50	5	
Bentazon								6					80		
DCPMU		14	17	8	8	5	6				3		11	4	6
Hydroxysimazine										16			96		

Chemical name	Type	Residential watersheds						Mixed-use watersheds				Agricultural watersheds			
		Kawa (ng/L)		Kaneohe (ng/L)		Manoa (ng/L)		Waimanalo (ng/L)		Kahawai (ng/L)		Honouliuli (ng/L)		Waikele (ng/L)	
		Base	Strom	Base	Storm	Base	Storm	Base	Storm	Base	Storm	Base	Storm	Base	Storm
Dietdrin	Insecticides	186	94	28	25	54	35								
Imidacloprid		46		11	47	16	51		45	13	54		264		31
Diazinon				10		21	25							23	105
Carbaryl			7				5		6		11				
Malathion		6		10	6				17				32		
Dinotefuran				4		10	4	3		29	6				3
Methoxyfenozide									4				85		
Fipronil		10	55	34	17	8	11		3						12
Fipronil amide		37	14	19	26	30	23			24	10				
Fipronil sulfone		8	9	14	6	4	8								8
Desulfinylfipronil		7	8	6	5		2								6
Azoxystrobin	Fungicides										2		58		2
Propiconazole		6		21		4			5		10		141		8
Carbendazim		29		16	16	9	12								10
4-Hydroxychlorothalonil				158	59								1,514		

Table 3.
Average herbicide, insecticide, and fungicide concentrations measured in Oahu streams, 2015 to 2025.

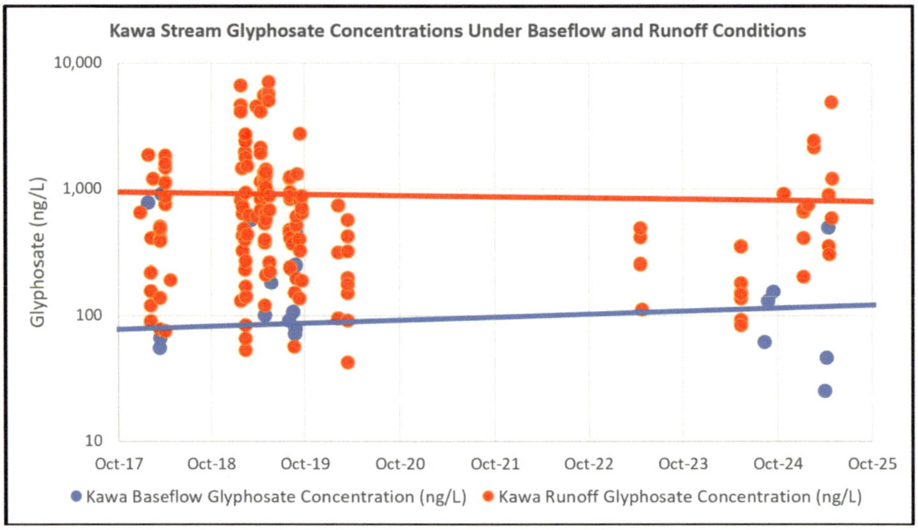

Figure 11.
Glyphosate concentration in Kawa Stream: 2017–2025.

5. Solvents

Industrial activities on Oahu are concentrated near the coast in areas underlain by nonpotable groundwater. It is in these coastal where most solvent contamination of shallow soils and groundwater occurs, typically in relation to dry-cleaning, degreasing, and automotive repair activities. The solvents trichloroethylene (TCE) and tetrachloroethylene (PCE), have been detected in a few potable water wells located in Central Oahu since the late 1980s, and the concentration levels in these deep wells continue to rise (**Figure 12**). This trend of rising solvent levels contrasts with the relatively rapid degradation of these same solvents observed in current and former industrial sites in the coastal areas of Oahu.

6. Pharmaceuticals

Between 2020 and 2022, researchers measured pharmaceutical and nutrient concentrations in perennial streams and springs across Oahu, analyzing over 700 samples from 36 streams and 11 springs situated in areas with varying densities of On-Site Disposal Systems (OSDS) and sewage lines [5]. The results indicated no statistically significant difference in pharmaceutical and nutrient levels between watershed with high or low densities of OSDS and sewer infrastructure. Under baseflow conditions, the average concentrations of caffeine, carbamazepine, and sulfamethoxazole in Oahu streams were 18%, 1.4% and 0.6% of average concentration levels in streams worldwide [26].

Figure 13 illustrates the spatial distribution of caffeine concentrations in streams and springs, revealing that the highest levels occurred in areas with low coastal groundwater fluxes. This pattern suggests that in regions with high

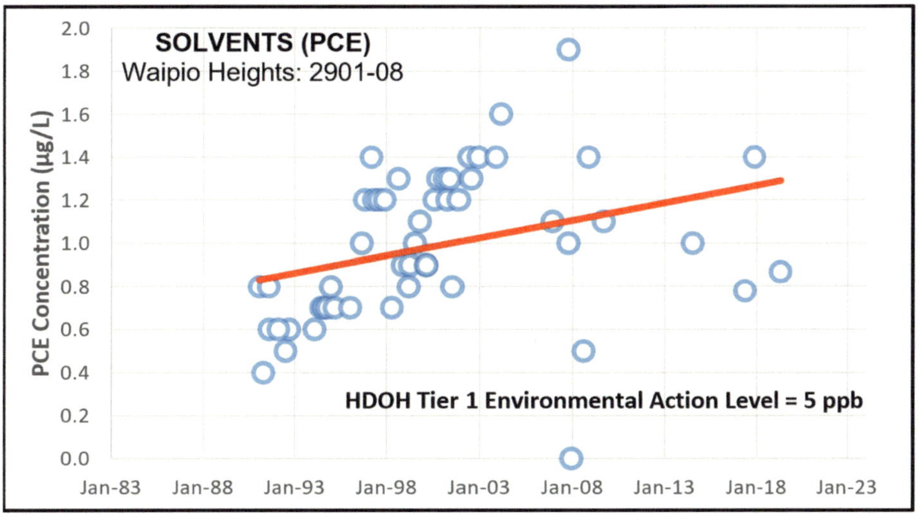

Figure 12.
Solvent (PCE) measured in Central Oahu wells.

underlying groundwater movement, wastewater inputs from cesspools and sewage lines are effectively diluted, resulting in minimal detectable concentrations [5].

Eight streams (five in leeward Oahu and three in windward Oahu) were sampled again under runoff conditions between 2023 and 2025. The average carbamazepine and caffeine levels in these eight streams increased from 2 and 469 PPT under baseflow conditions to 13 and 715 PPT under runoff conditions (~650% and 155% increase, respectively), based on 200 individual analyses. However, even under runoff conditions, the carbamazepine and caffeine concentrations measured in Oahu streams are 15% and 47% of the average concentration levels of these two compounds measured in over 1,200 streams sampled around the world [26]. This wastewater tracer data collected suggests that the flux of wastewater from cesspools and leaky sewers to Oahu streams is minimal under dry conditions but does increase somewhat under rainfall runoff conditions.

Table 4 presents the pharmaceutical and ammonia concentrations measured in streams located in residential and mixed-use watersheds under baseflow and runoff conditions. The average concentrations of these constituents, measured in wastewater collected from large public septic tanks and at wastewater treatment plants [13] around Oahu, are also compiled in this table.

7. Major and trace metals

Figure 14 shows the median concentration of major and trace elements measured in 10 Oahu streams located in nonagricultural areas under groundwater-dominant drought conditions. Sodium, silica, magnesium, calcium, and potassium are present in Oahu streams at milligram per liter (mg/L) concentration levels. Chloride and sulfate are also present at mg/L levels in Oahu streams under baseflow conditions. Iron, strontium, manganese, vanadium, and barium are present in Oahu streams

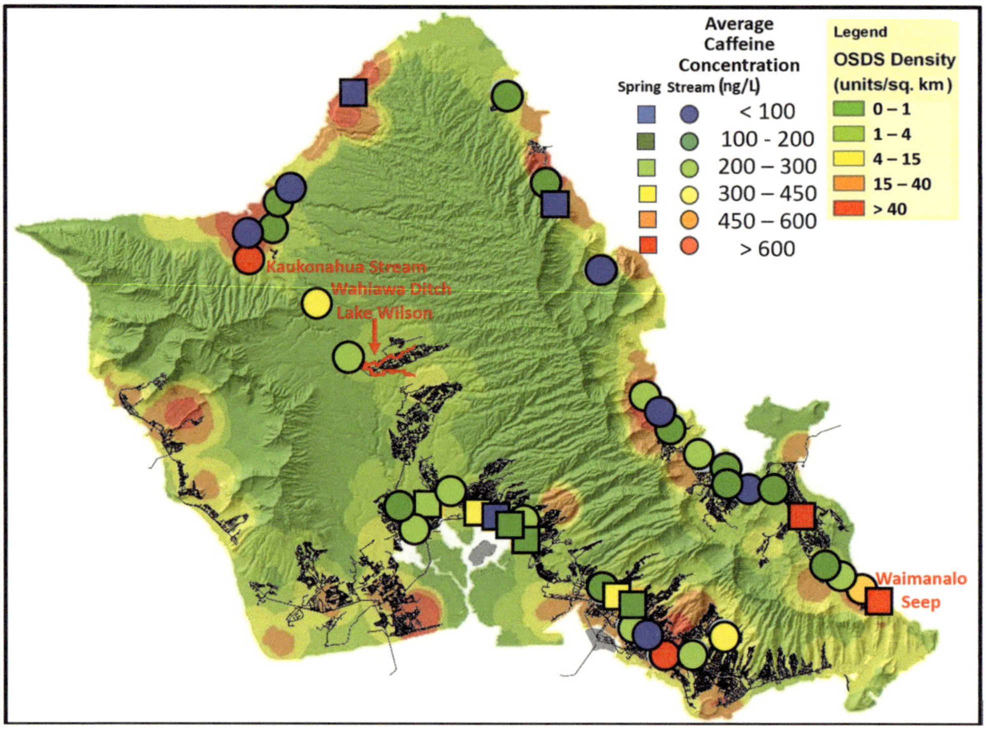

Figure 13.
Caffeine concentrations in streams and springs on Oahu.

at µg/L concentration levels. Zinc, nickel, selenium, copper, chromium, molybdenum, cobalt, boron, arsenic, and lead are present in Oahu streams at ng/L concentration levels.

Table 1 shows the significant increase in metals and nutrients observed in a stream that receives input from groundwater impacted by historical agricultural use of the overlying lands. The increase in silica concentration in wells located within and downgradient of agricultural fields was first noted in the 1960s [27]. Groundwater concentration levels above 40 mg/L silica are associated with areas on Oahu affected by return irrigation water [5]. Vanadium concentrations in groundwater underlying former agricultural fields are also particularly elevated. This is likely related to the historic furrow irrigation of sugarcane fields, which leached vanadium and silica from shallow soils. **Figure 15** illustrates the spatial distribution of vanadium concentrations measured in groundwater and streams on Oahu. Groundwater and streams that have not been impacted by historical agricultural activities generally exhibit vanadium concentrations of less than 15 µg/L.

A recent epidemiological study on Oahu found a correlation between the concentration of vanadium in groundwater and the prevalence of NTM pulmonary disease [28]. The occurrence of NTM is increasing in the United States and Hawaii has the highest prevalence of NTM disease in the country. This disease is characterized by nonspecific symptoms including chronic cough and fatigue. The concentration levels of vanadium in Central Oahu wells are generally three to four times higher than the background vanadium levels present in groundwater

Chemical name	Unit	Residential watersheds						Mixed-use watersheds						Average wastewater
		Kawa		Makiki		Manoa		Waimano		Waimanalo		Kahawai		
		Base	Storm	Base	Storm	Base	Storm	Base	Storm	Base	Storm	Base	Storm	
Caffeine	ng/L	341	459	372	774	241	1,102	372	676	452	732	259	886	**36,930**
Carbamazepine	ng/L	2.0	7.4	2.9	17.6	0.2	16.3	2.9	11.8	5.0	12.9	0.0	3.4	**503**
Ammonia	mg/L	0.04	0.06	0.12	0.17	0.09	0.08	0.06	0.09	0.07	0.16	0.13	0.12	**21**

Table 4.
Mean Pharmaceutical levels measured in Oahu streams and wastewater.

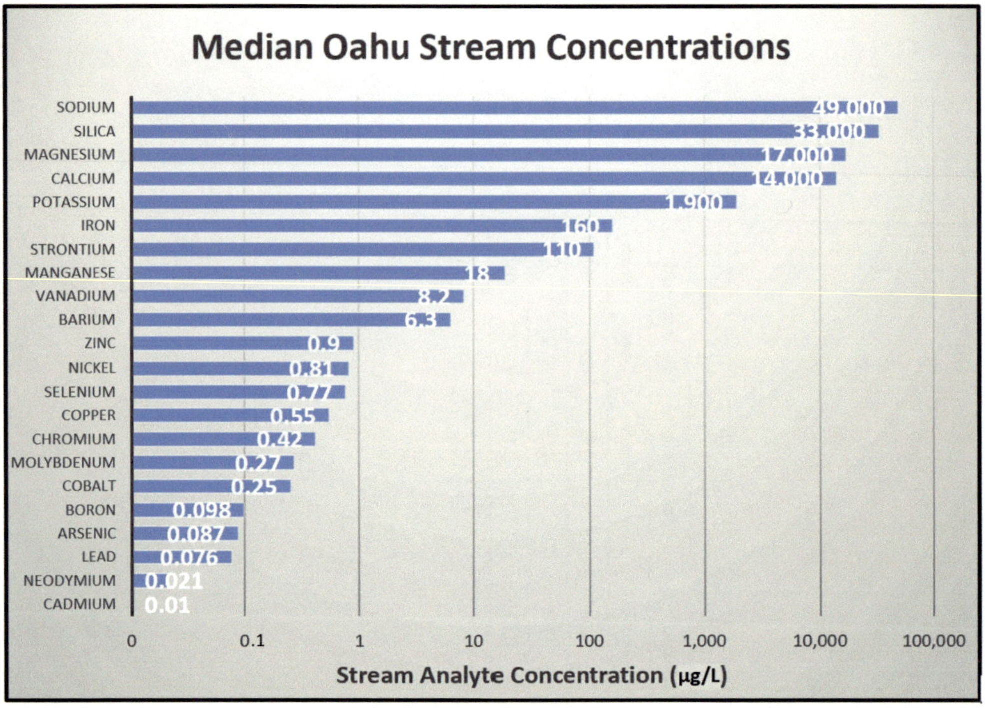

Figure 14.
Major and trace element concentrations in Oahu streams.

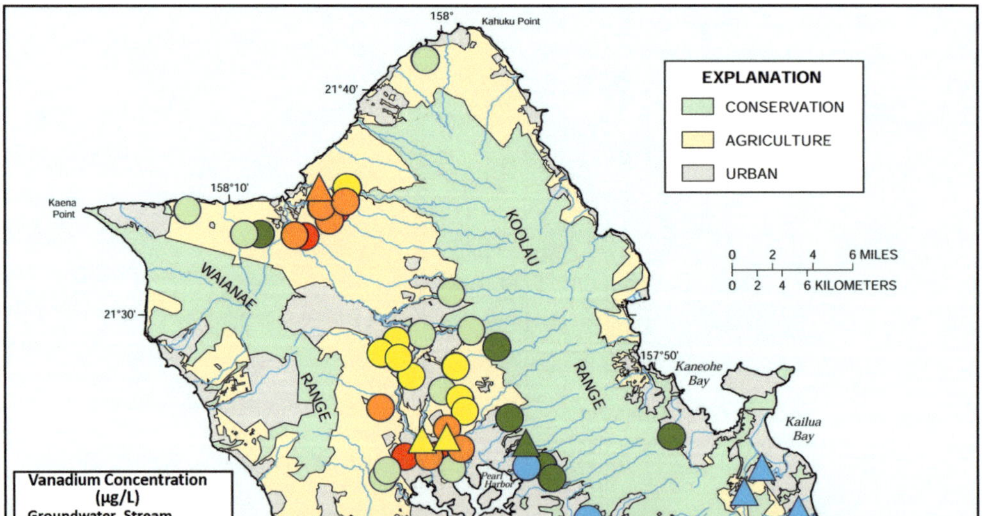

Figure 15.
Vanadium concentrations in groundwater and streams on Oahu.

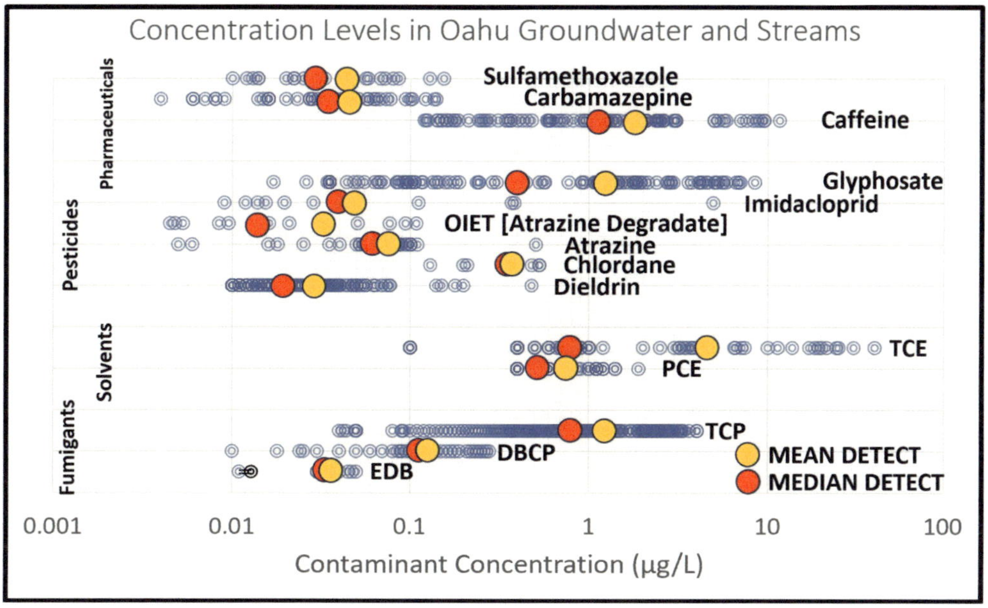

Figure 16.
Range of contaminant concentrations detected on Oahu.

elsewhere on Oahu (**Figure 15**). The epidemiological study found that increased concentrations of vanadium in drinking water in Central Oahu were associated with a statistically significant increased risk of Mycobacterium avium complex infection [28]. Specifically, island residents who receive drinking water from wells in Central Oahu containing elevated levels of vanadium had statistically higher rates of NTM infection than the rest of the population on Oahu.

8. Conclusions

The concentration levels of nutrients detected in Oahu streams suggest that the current State of Hawaii's WQSs require an upward revision. The concentration levels of nonglyphosate herbicides, fungicides, and insecticides detected in Oahu streams are generally below 60 ng/L, while glyphosate concentrations were generally detected at high ng/L to low µg/L levels. Despite the discontinued use of glyphosate in weed-killing products sold for residential use in 2023–2024, the measured levels of glyphosate exiting a residential watershed in windward Oahu (Kawa) have yet to decline. In addition, the concentration levels of the insecticide dieldrin and soil fumigants DCBP and TCP continue to rise in potable aquifers and streams (dieldrin) on Oahu, despite the use of these chemicals being halted in the 1980s. The elevated concentration levels of vanadium in Central Oahu wells have been linked to an elevated incidence of pulmonary disease in residents serviced by these wells. On a positive note, little evidence has been found for large-scale groundwater or stream contamination on Oahu from OSDS and sewage conveyance system exfiltration. In addition, many of the pesticides that were formerly used to cultivate sugarcane, such as atrazine, have declined to nondetectable or low part-per-trillion levels. **Figure 16** summarizes the range of concentrations and the mean and median concentrations of fumigants, solvents, pesticides, and pharmaceutical compounds measured in Oahu groundwater and streams in environmental studies conducted since 1989.

Author details

Steven R. Spengler and Samuel C. Spengler
Pacific Rim Water Resources, Hawaii, USA

*Address all correspondence to: steve.spengler@gmail.com

References

[1] Anthony SS, Hunt Jr. CD, Brasher AM, Miller LD, Tomlinson MS. Water Quality on the Island of Oahu. Vol. **1239**. Hawaii: Reston, Va., U.S. Geological Survey Circular; 2004. p. 37

[2] Bevenue A, Hylin JW, Kawano Y, Kelley TW. Organochlorine pesticide residues in water, sediment, algae, and fish, Hawaii—1970–1971. Pesticide Monitoring Journal. 1972;**6**(1):56–64

[3] State of Hawaii, 2025. The State of Hawaii data book 2023, a statistical abstract: State of Hawaii, Department of Business, Economic Development, and Tourism. Available from: http://www.hawaii.gov/dbedt/db00 [Accessed: 2025-August-2]

[4] Kāne HL, Mair A, Mifflin J, 2024, Mean annual water-budget components for Kaua'i, O'ahu, Moloka'i, Lāna'i, Maui, and the Island of Hawai'i, for a set of recent and future-climate conditions, and 2020 land cover: U.S. Geological Survey data release. DOI: 10.5066/P972KMSL

[5] Spengler SR, Heskett MD. Impact to stream water quality from sewage exfiltration and legacy on-site disposal systems on the island of O'ahu, Hawaii. International Journal of Environmental Impacts. 2023;**6**(1):13–23

[6] Lau LS. 1973. Quality of Coastal Waters, Second Annual Progress Report. Water Resources Research Center, Technical Report No. 77, September 1973

[7] Oki DS, Giambelluca TW.DBCP, EDB, and TCP Contamination of Ground Water in Hawaii. Groundwater. 1987;**25**(6):693–702

[8] Hawaii Department of Health (HDOH) Hazard Evaluation and Emergency Response Office. 2013–14 state wide pesticide sampling pilot project water quality findings. A Joint Investigation by the Hawaii State Departments of Health and Agriculture. 2014;**2014**:26

[9] Spengler SR, Heskett MD, Gray JI. Pesticide levels in streams and sediments on the islands of Oahu and Kauai, Hawaii. International Journal of Environmental Impacts. 2019;**10** (9):1–17

[10] Johnson AG, Kennedy JJ, and Alvarez DA. 2024. Pesticides in surface water downstream of and near agricultural and developed land in Hawai'i, 2015–19: U.S. Geological Survey Scientific Investigations. Report 2024–5071. 94 p. DOI: 10.3133/sir20245071 [Accessed: 2025-August-15]

[11] Spengler SR, Heskett MD, Spengler SC. Glyphosate in runoff from urban, mixed-use and agricultural watersheds in Hawaii, USA. WIT Transactions on Ecology and the Environment. 2018;**228**. DOI:10.2495/WP180081

[12] Spengler SR, Heskett MD, Spengler SC. Occurrence and persistence of the herbicide glyphosate in a suburban tropical watershed. International Journal of Environmental Impacts: Management, Mitigation and Recovery. 2021;**4**(2):141–156

[13] Hawaii Department of Health. 2019. 2018 Groundwater Status Report. Report Prepared for the Groundwater Protection Program dated October 21, 2019

[14] Spengler SR, Heskett MD. Identification of sewage exfiltration in coastal areas through the monitoring of drugs and stimulant concentrations in urban storm drains. Urban Water Systems and Floods. 2022;**IV**:67–80

[15] Voss CI, Wood WW. 1993. Synthesis of geochemical, isotopic and groundwater modelling analysis to explain regional flow in a coastal aquifer of southern Oahu, Hawaii. IAEA-TECDOC-777, Proceedings of a Final Research Coordination Meeting held in Vienna, 1-4 June 1993

[16] WSP, 2019. Oahu water quality modeling analysis. City and County of Honolulu Department of Facility Maintenance Storm Water Quality Branch. Final Report. dated September 2019

[17] Hawaii Department of Health. 2024. 2024 state of Hawaii water quality monitoring and assessment report: Integrated report prepared by the Clean Water Branch to the U.S. Environmental Protection Agency. dated January 8, 2024

[18] Young RHF, Lau LS, Konno SK, Oldnall RJ, Lee HH. Water quality monitoring: Kaneohe Bay and selected watersheds, July to December 1975. Water Resources Research Center, Technical Report No. 98. May 1976

[19] Lee CMY. 2012. Chlordane and Dieldrin Contamination and Biodegradation in Ala Wai Canal Sediment. Master's Thesis in Civil Engineering. University of Hawaii

[20] Mink JF 1981. DBCP and EDB in soil and water at Kunia, Oahu, Hawaii. Report Prepared for Del Monte Corporation. Honolulu, Hawaii

[21] Lau LS. 1987. Organic chemical contamination of O'ahu groundwater. Tech. Rep. No. 181. Water Resources Research Center, University of Hawaii

[22] Or S, Lau LS 1987. Trace organic (DBCP) transport simulation of Pearl Harbor aquifer, O'ahu, Hawaii: Multiple mixing-cell model, phase 1. Tech. Rep. No. 174. Water Resources Research Center, University of Hawaii

[23] Hunter CL, Stephenson MD, Tjeerdemas RS, Crosby DG, Ichikawa GS, Goetzl JD, Paulson KS, Crane DB, Martini M, Newmans JW. Contaminants in oysters in Kaneohe Bay, Hawaii. Marine Pollution Bulletin. 1995;**30**(10):646–654

[24] Sava V, Velasquez A, Song S, Sanchez-Ramos J. Dieldrin elicits a widespread DNA repair and antioxidative response in mouse brain. Journal of Biochemical and Molecular Toxicology. 2007;**21** (3):125–135

[25] Maggi F, La Cecilia D, Tang FHM, McBratney A. The Global Environmental Hazard of Glyphosate Use. Science of the Total Environment. 2020;**717**:137167

[26] Wilkenson JL, et al. Pharmaceutical pollution of the world's rivers. Proceedings of the National Academy of Sciences. 2022;**119**(8): e2113947119. DOI: 10.1073/pnas.2113947119

[27] Davis SN 1969. Silica in streams and ground water in Hawai'i. Water Resources Research Center Technical Report No. 20. January 1969

[28] Lipner EM, French JP, Nelson S, Falkinham JO, Marcaldo RA, Blakney

RA, Daida YG, Frankland TB, Messier KP, Honda JR, Honda S, Prevots DR. Vanadium in groundwater aquifers increases the risk of MAC pulmonary infection in O'ahu, Hawaii. Environmental Epidemiology. 2022;**6**:e220

Chapter 5

Environmental Perspectives on Biopesticides: Current Knowledge and Research Gaps

Derya Kocamaz and Oguz Merhan

Abstract

Biopesticides have emerged as promising alternatives to synthetic pesticides, especially in response to growing concerns about environmental pollution, risks to non-target organisms, and potential threats to human health. Derived from natural sources such as plants, microorganisms, and biopolymers, they offer several advantages, including high target specificity, environmental safety, and minimal chemical residues. Despite these benefits, their practical application is often challenged by limited stability, short shelf life, and variable field performance under different environmental conditions. The adoption of green synthesis strategies – utilizing renewable resources, non-toxic solvents, and energy-efficient processes – has improved the formulation stability and ecological compatibility of these products. However, environmental data on biopesticides remain limited, and their long-term impacts on non-target organisms and ecosystems are not yet fully understood. This chapter provides a comprehensive overview of the current state of biopesticide development, their mechanisms of action, formulation improvements through green synthesis, and the existing knowledge gaps in their environmental evaluation. Emphasis is placed on the need for multidisciplinary approaches to ensure their safety, stability, and effective integration into sustainable pest management systems.

Keywords: biopesticides, green synthesis, toxicology, environmental impact, public health, risk assessment

1. Introduction

Pesticides are natural or synthetic chemical compounds used to prevent, control, and eliminate insects, weeds, and other harmful organisms that adversely affect plant growth [1, 2]. Since the mid-1940s, particularly with the expansion of commercial agriculture, there has been a significant and continuous increase in global pesticide demand [3, 4]. By preserving crop yield and quality, pesticides make a substantial contribution to food security. Moreover, in public health, they are effectively used for controlling vectors such as ticks, mosquitoes, and fleas, thereby preventing the spread of infectious diseases, including malaria, dengue fever, and Lyme disease [5, 6].

Although their role in enhancing agricultural productivity is well documented, the excessive and uncontrolled use of pesticides has led to severe environmental and health problems. These include food contamination, soil and water pollution, ecosystem disruption, and biodiversity loss [7, 8]. Most pesticides are not species-specific and may exert toxic effects on non-target organisms, such as beneficial insects, pollinators, aquatic species, and even humans. Studies have consistently reported pesticide residues in fruits, vegetables, processed food, water, air, and soil. In particular, acute and chronic health issues associated with pesticide exposure remain a major public health concern in developing countries. Furthermore, the emergence of pesticide resistance, along with their potential genotoxic, neurotoxic, and endocrine-disrupting properties, has raised serious concerns regarding the long-term sustainability of synthetic pesticide use [9, 10]. Previous studies have highlighted that these chemicals may induce carcinogenic, cytotoxic, and mutagenic effects in humans [11–15].

Against this background, the development and application of novel pesticides that are non-toxic to humans, animals, and the environment have become a research priority. As a result of this transition, biopesticides have emerged as promising alternatives to synthetic pesticides, offering several advantages in pest management. Biopesticides are characterized by low-cost production, eco-friendly nature, high specificity against target organisms, absence of harmful residues, and overall sustainability [16]. The rapidly growing biopesticide market further indicates that these products are becoming essential components of modern agriculture [17].

Despite the reported advantages of biopesticides, their limitations should not be overlooked. It has been reported that field performance varies when applied under different climate and soil conditions, and their short shelf life and low storage stability pose significant challenges for users [18, 19]. Additionally, in the case of botanical biopesticides, seasonal fluctuations in the concentrations of bioactive metabolites complicate standardization and result in inconsistent efficacy [20]. Moreover, from the perspective of farmers and users, the slower mode of action and efficacy duration undermine trust and reduce the perceived benefits for producers [21]. For example, although botanical biopesticides demonstrate strong pesticidal activity, their short persistence under field conditions has limited their commercial availability to only a few volatile oil-based formulations [22–25]. Similarly, microbial pesticides are often specific to only one pest species and are highly sensitive to environmental conditions such as heat, temperature, and UV radiation, which can decrease their activity and result in crop damage in practical use [26, 27]. Although these examples demonstrate that biopesticides may not fully replace synthetic pesticides in a simple "use-and-replace" model, encapsulation technologies such as nano- and microencapsulation have enabled improvements in shelf life and stability. These developments have contributed to the increasing adoption of green synthesis principles in biopesticide production processes [28, 29]. Green synthesis is an approach designed to reduce the environmental impact of chemical synthesis and production. It includes principles such as the use of non-toxic solvents, prioritization of renewable raw materials, waste minimization, energy efficiency, and environmental safety [30]. The integration of this approach into biopesticide production not only reduces the ecological footprint of the process but is also believed to positively affect the safety profile of the final products [31]. Despite these developments, a critical question remains: To what extent can we trust the safety of biopesticides produced through green synthesis in terms of human health and environmental impact? Answering this question requires moving beyond the common assumption that "natural is safe" and instead emphasizes the necessity of rigorous, evidence-based toxicological evaluations. This is because the design choices that make green synthesis attractive may also influence the toxicity profiles of biopesticides. For example, plant polyphenols used as reducing/capping agents, biopolymers such as chitosan and starch used as carriers, and metals such as Ag, CuO, and ZnO alter ion release, surface reactivity, and persistence of biopesticides [32, 33]. In nanobiopesticides, particle size, surface charge, uptake, and distribution affect their interactions with biological membranes and microbiomes. This indicates that green-synthesized biopesticides are not necessarily less toxic and that their toxicity profile depends on their composition, size, coating, frequency of use, and dosage. For example, nanoparticles smaller than 20 nm may accumulate in biological systems and pose environmental risks [34, 35]. Similarly, non-biodegradable particles such as silica can persist in soil and aquatic ecosystems [11]. The limited number of studies on the potential accumulation and degradation time of biopesticides containing metals such as Ag, CuO, and ZnO [36, 37], as well as the unknown long-term effects on soil microorganisms and symbiotic fungi, may negatively affect soil microbiota and nutrient cycling.

Increasing public trust in these next-generation agricultural agents depends on the comprehensive characterization of their toxicological profiles, acknowledgment of existing knowledge gaps, and the development of a clear research roadmap for future investigations. Today, the safety of biopesticides is assessed through multi-tiered toxicity testing in accordance with international standards, particularly those defined by the Organisation for Economic Co-operation and Development, the Environmental Protection Agency, and the European Food Safety Authority (OECD, EPA, and EFSA). This system includes parameters such as acute toxicity, pathogenicity, environmental fate, and effects on non-target organisms. Acute tests (e.g., OECD 201, 202, 203) are applied to fish, *Daphnia magna*, algae, and earthworms, and ecotoxicological impacts are also evaluated in birds, bees, and plants. In addition, under next-generation approaches, genomics, proteomics, metabolomics, and ecological modeling are used to understand the mechanistic toxicity profiles of biopesticides.

This assessment process is not limited to *in vivo* and *in vitro* tests but also includes bioinformatic off-target effect modeling and adverse outcome pathway-based strategies, particularly for novel agents such as RNAi-based biopesticides [38]. Through omics-based evaluations, the molecular-level toxic responses, off-target effects, and environmental interactions of biopesticides can be elucidated [39]. As a result, the environmental safety of biopesticides is evaluated holistically, encompassing both short-term and long-term effects [40].

2. Biopesticides: Classification, sources, and mechanisms of action

Biopesticides are generally classified into three groups based on their source and production method: phytopesticides (plant-derived), microbial pesticides (microorganism-derived), and nanobiopesticides (biological nanoparticles). In recent years, additional categories, such as RNAi-based biopesticides, algal/cyanobacterial agents, and aquatic macrophyte-derived pesticides, have emerged, reflecting the growing diversity and complexity of biopesticidal technologies [41].

2.1 Phytopesticides

Phytopesticides are natural pesticides obtained from different plant organs such as leaves, roots, stems, and seeds. These products can be prepared in the form of powders, extracts, essential oils, aqueous solutions, or suspensions from plant-derived raw materials and are widely used in the control of plant pathogens and harmful organisms [42–44]. To date, more than 2,000 plant species have been reported to possess insecticidal properties, with families such as *Myrtaceae, Lauraceae, Rutaceae, Lamiaceae, Asteraceae, Apiaceae, Cupressaceae, Poaceae, Zingiberaceae, Piperaceae, Liliaceae, Apocynaceae, Solanaceae, Caesalpinaceae,* and *Sapotaceae* standing out due to their abundance of bioactive compounds [45].

The major bioactive constituents of phytopesticides are secondary metabolites, including steroids, alkaloids, tannins, phenolics, flavonoids, and resins – which are regarded as the principal active agents underlying their pesticidal effects [46]. In comparison with synthetic insecticides, phytopesticides are considered more environmentally friendly, less toxic to non-target organisms, and associated with a lower risk of resistance development. However, their rapid degradation under environmental factors such as sunlight, humidity, and wind requires their application closer to harvest. Moreover, seasonal fluctuations in the concentration of active compounds hinder standardization, and their poor storage stability further limits the number of phytopesticides that have been successfully commercialized and registered [18].

The modes of action of phytopesticides against pests are diverse. Plant-derived insecticides act through different biological mechanisms; some function as repellents by deterring pests from host plants, whereas others act as attractants by luring pests into specific areas. In addition, these compounds can disrupt pest reproduction by directly inactivating eggs and larvae, thereby restricting population growth. Importantly, phytopesticides exert their effects not only through direct insecticidal activity but also via repellent, antifeedant (feeding deterrence), and growth or reproduction – inhibiting properties, offering significant potential for their integration into sustainable pest management strategies [31].

2.2 Microbial biopesticides

Microbial biopesticides, derived from microorganisms such as bacteria, fungi, viruses, and protozoa, represent sustainable and environmentally friendly alternatives to conventional chemical pesticides [47]. They are currently available in diverse formulations, including wettable powders, aqueous suspensions, oil-based dispersions, water-dispersible granules, and suspension seed coatings [48]. The antagonistic activity of microbial biopesticides against plant pathogens and pests is mediated through multiple mechanisms, such as toxin production, enzyme secretion, volatile compound release, direct colonization or consumption of the host, induction of systemic resistance in crop plants, and competition for nutrients and ecological niches [22]. Owing to these attributes, microbial biopesticides not only protect ecosystems but also play a pivotal role in integrated pest management (IPM) strategies designed to maintain high crop productivity. Recent evidence further indicates that microbial biopesticides enhance soil health and biodiversity, thereby aligning modern agricultural practices with broader ecological and sustainability objectives [49]. Historically, the agricultural application of beneficial microorganisms was restricted to a few well-characterized taxa, such as mycorrhizal fungi and *Rhizobium* bacteria, whose plant growth-promoting mechanisms were extensively studied [50]. In recent decades, however, the diversity of microorganisms exploited as biopesticides has expanded significantly. Representative taxa include bacterial genera such as *Chromobacterium, Pseudomonas, Bacillus,* and *Yersinia*; fungal genera including *Beauveria, Paecilomyces, Verticillium, Hirsutella, Metarhizium,* and *Lecanicillium*; as well as entomopathogenic nematodes (EPNs) from the genera *Steinernema* and *Heterorhabditis* [41]. Beyond pest suppression, microorganisms employed as biopesticides have been shown to exhibit multiple plant growth-promoting traits. These include phosphate solubilization, nitrogen fixation, 1-aminocyclopropane-1-carboxylate deaminase activity, siderophore production, biofilm formation, phytohormone synthesis, and enhancement of tolerance or resistance to biotic and abiotic stresses [51, 52]. Collectively, these multifunctional properties underscore the dual role of microbial biopesticides in crop protection and sustainable agricultural development.

Moreover, microalgae and cyanobacteria – such as *Chlorella vulgaris*, *Anabaena flos-aquae*, and *Nostoc piscinale* – have demonstrated antifungal, antibacterial, and insecticidal effects and are now recognized as emerging sources of biopesticides. Their ability to rapidly grow in controlled environments and synthesize diverse allelochemicals makes them especially promising in aquatic and closed-system agriculture. However, their commercialization is still limited by scalability, stability, and formulation constraints [53, 54].

2.3 Nanobiopesticides

Nanobiopesticides have emerged through the integration of nanomaterials with biopesticides to overcome key limitations of conventional biopesticides, such as thermal instability, poor persistence, low efficacy, and high susceptibility to environmental variations compared to chemical pesticides. Due to their intrinsic nanoscale dimensions, nanomaterials exhibit advantageous physicochemical properties – including larger surface-to-volume ratio, enhanced solubility, permeability, and overall reactivity – relative to bulk materials. These features confer biopesticides with improved stability, bioavailability, efficacy, adhesion, target specificity, environmental safety, and controlled release of the active ingredient [55, 56]. The major categories of nanomaterials, based on their composition, include pure carbon nanostructures, inorganic nanomaterials, organic nanomaterials, and organic–inorganic hybrids. Among the most widely used nanoparticles in agriculture are silver, copper, titanium, zinc, silica, gold, aluminum, and iron nanoparticles, as well as chitin nanoparticles, chitosan, nanoclay, graphene nanoparticles, multi-walled carbon nanotubes, and biopolymers [57–60]. Nanopesticides, generally produced and applied within the nanoscale range of 1–100 nm, are classified according to their delivery forms as nanospheres, nanocapsules, nanoemulsions, nanogels, liposomes, and inorganic nanocarriers [37].

2.4 RNA interference (RNAi)-based biopesticides

Increasing resistance of pests and pathogens to conventional pesticides, along with the limitations of existing biopesticides, has led to the incorporation of RNAi technology into biopesticide development. This technology has been successfully employed in transgenic crops such as maize, soybean, and cotton to enhance resistance against major insect pests [41]. The RNAi mechanism relies on the expression of transgenic double-stranded RNA (dsRNA), which induces virus resistance and gene silencing in plants. In this process, "guide RNAs," approximately 21–25 nucleotides in length, are generated as intermediates; these guide RNAs specifically direct the degradation of target RNAs [61–63], thereby blocking essential physiological or developmental pathways of the pest organism and resulting in mortality or developmental arrest. RNAi-based biopesticides can be delivered not only through transgenic plants but also as exogenous applications, such as foliar sprays or seed treatments. For example, dsRNA formulations have demonstrated high efficacy against key agricultural pests, including *Diabrotica virgifera* (Western corn rootworm) and *Helicoverpa armigera* (cotton bollworm) [64, 65]. In June 2017, the US EPA approved SmartStax PRO, the first RNAi-based plant-incorporated protectant, to help US farmers control Western corn rootworm – a devastating pest that had developed resistance to multiple other pesticides [66]. The key advantage of RNAi-based biopesticides is their high specificity and minimal off-target effects. However, several challenges remain unresolved, including dsRNA degradation under environmental conditions, low uptake efficiency by target pests, stability during field application, and production costs. Furthermore, regulatory frameworks for RNAi products are still evolving, and standardized protocols for assessing environmental fate, persistence, ecotoxicity, and effects on non-target organisms are urgently needed before large-scale commercialization can be achieved [67, 68].

2.5 Aquatic macrophyte-derived biopesticides

Aquatic macrophytes have gained increasing attention as promising sources of biopesticidal compounds due to their allelopathic potential. Species such as duckweed (*Lemna minor*), muskgrass (*Chara* spp.), water hyacinth (*Eichhornia crassipes*), hydrilla (*Hydrilla verticillata*), water lettuce (*Pistia stratiotes*), and filamentous algae (*Lyngbya wollei*) have been reported to produce secondary metabolites capable of inhibiting the growth, development, survival, and reproduction of surrounding organisms [69–71]. These allelopathic compounds interfere with vital biological functions in target organisms and have shown activity against insects, weeds, and pathogens. For instance, while neem (*Azadirachta indica*) extract is well known for its broad-spectrum insecticidal activity, *Eichhornia crassipes* has demonstrated the ability to suppress the development of *Spodoptera litura*, a lepidopteran pest of economic importance [71, 72]. Similarly, the weed species *Chenopodium album* has been shown to exhibit suppressed growth in the presence of duckweed and water lettuce, further supporting the allelopathic potential of these aquatic plants [41, 71]. Aquatic macrophytes produce bioactive allelochemicals with inhibitory effects on plant pathogens; however, their translation into commercially viable biopesticide products is still in its infancy.

2.6 Modes of action of biopesticides

Biopesticides exert their biological effects through five principal modes of action: phytopesticidal, microbial, nanobiopesticidal, RNAi-based, and aquatic macrophyte-derived actions. Each mode is characterized by specific molecular and physiological mechanisms that interfere with the survival, growth, or reproduction of the target organisms.

2.6.1 Phytopesticides

Plant-derived biopesticides exhibit multiple modes of action, functioning primarily as herbicidal, insecticidal, and fungicidal agents through interference with essential biological pathways in target organisms. Their herbicidal activity is often attributed to allelochemicals – such as monoterpenes and sesquiterpenes – released via allelopathic interactions. Essential oils (PEOs) from *Artemisia* spp. and *Xanthium strumarium L.*, containing bioactive compounds including artemisia ketone, 1,8-cineole, and yomogi alcohol, have been shown to inhibit the germination of *Amaranthus retroflexus* and *Setaria viridis* at concentrations as low as 10–1,000 mg L^{-1} [73]. Similarly, PEOs extracted from *Pinus radiata* and *Cupressus sempervirens*, rich in α-pinene, δ-3-carene, and germacrene D, exhibit phytotoxicity against *Sinapis arvensis*, *Lolium rigidum*, and *Phalaris canariensis* by reducing α-amylase activity and disrupting pigment biosynthesis during early plant development [74–76]. Compounds such as sarmentin, isolated from *Piper* spp., have been reported to inhibit Photosystem II by competitively binding to plastoquinone sites, thereby disrupting the photosynthetic electron transport chain.

In terms of insecticidal activity, plant essential oils exert neurotoxic effects by targeting key neuronal enzymes and receptors. For example, oils derived from *Citrus grandis* have demonstrated larvicidal and adulticidal effects, attributed to the inhibition of acetylcholinesterase, modulation of GABA-gated chloride channels, and interference with cellular respiration and membrane stability [77–79]. Monoterpenes in these oils also induce cytotoxicity by impairing mitochondrial integrity and reducing membrane permeability.

Fungicidal effects are associated with structural damage to fungal hyphae and organelles. Treatment of *Aspergillus niger* with essential oils from *Thymus eriocalyx* and *Thymus x-porlock* has led to mitochondrial disruption, cytoplasmic coagulation, and compromised cell wall and membrane integrity [80].

2.6.2 Microbial biopesticides

The insecticidal efficacy of microbial biopesticides is based on a range of mechanisms that target fundamental physiological pathways, such as membrane disruption, enzymatic degradation of protective barriers, systemic infection, and interference with metabolic processes. Among these, the most well-known mechanism involves the crystalline δ -endotoxins (Cry and Cyt proteins) produced by *Bacillus thuringiensis* (Bt) during sporulation. When susceptible larvae ingest these toxins, the proteins are solubilized in the alkaline environment of the midgut, activated by proteases, and bind to specific receptors on epithelial cells. This interaction leads to pore formation, osmotic imbalance, and cell lysis, resulting in paralysis and death within a short period [81]. Other bacterial species, such as *Bacillus sphaericus,* produce specific toxins such as the binary toxin that exhibit potent larvicidal activity against mosquitoes, whereas biocontrol agents such as *Bacillus subtilis* secrete hydrolytic enzymes, such as chitinases and proteases, along with secondary metabolites, to disrupt cuticular integrity or interfere with digestive functions. In contrast, EPNs, belonging to the genera *Steinernema* and

Heterorhabditis [82], invade the insect host and release their symbiotic pathogenic bacteria (*Xenorhabdus* or *Photorhabdus* spp.), which rapidly cause septicemia and death within 48 hours. The nematodes then reproduce within the host cadaver, producing new infective juveniles (IJs) that ensure persistence and recycling of the biocontrol agent in the ecosystem. Finally, protozoan biopesticides, such as *Nosema locustae*, induce chronic infections by entering host cells after the ingestion of spores. Through intracellular proliferation, these microsporidia damage tissues and vital organs, ultimately impairing feeding, growth, and reproduction. Although the onset of pathogenic effects is slow, such infections provide long-term population suppression of pest species.

2.6.3 Nanobiopesticides

The mechanism of nanobiopesticides is primarily based on their ability to efficiently deliver bioactive molecules to the target site and disrupt the fundamental cellular functions of pests and pathogens. The development of various nanoscale carrier systems, including nanoemulsions, nanocapsules, nanocontainers, and nanocages, has significantly contributed to advances in biopesticide technology. These nanosystems are designed to enhance the protection, dispersion, and controlled release of active ingredients used in crop protection [83]. By improving the bioavailability and stability of the active compounds, they minimize environmental degradation and reduce loss during application. At the cellular level, cationic polymers integrated into nanobiopesticide formulations can electrostatically bind to the negatively charged microbial surfaces. This electrostatic interaction destabilizes the cell membrane, leading to leakage of intracellular contents and, ultimately, cell lysis and death. The biocidal effect is further enhanced by the presence of quaternary ammonium groups in some nanoparticle structures. These groups, which may be arranged as long amino side chains, can target a broad range of bacterial and fungal species by altering membrane permeability and damaging cellular integrity. Due to their strong biocidal properties and high activity under diverse environmental conditions, polymers containing quaternary ammonium groups are widely used in the design of nanobiopesticide systems. A wide range of functional polymers – including amphiphilic copolymers, functionalized cationic polycarbonates, poly(amidoamines), and polyethyleneimines – has been synthesized and evaluated for their pest-control potential [78–80]. These materials can act either as carriers that facilitate adhesion and penetration through insect cuticles and microbial biofilms, or as active biocidal agents that directly interact with cell membranes and metabolic pathways. In summary, nanobiopesticides exert their effects through a combination of electrostatic, oxidative, and biochemical mechanisms, leading to membrane disruption, enzyme inactivation, and inhibition of vital metabolic processes in target organisms [84].

2.6.4 RNAi-based biopesticides

The mode of action of RNAi-based biopesticides relies on the sequence-specific silencing of essential genes in target pests. These biopesticides contain dsRNA molecules designed to match the nucleotide sequence of a target messenger RNA. Once ingested or absorbed by the pest, the dsRNA is processed by the *Dicer* enzyme into small interfering RNAs (siRNAs). These siRNAs are incorporated into the RNA-induced silencing complex, which recognizes and binds complementary mRNA transcripts. The bound mRNA is then cleaved and degraded, resulting in translational inhibition and gene silencing [85]. By blocking the expression of vital genes responsible for growth, reproduction, or metabolism, RNAi triggers developmental arrest or mortality in pests. This mechanism provides exceptionally high specificity, as dsRNA molecules act only on sequences complementary to their target genes, minimizing non-target effects [86].

2.6.5 Aquatic macrophyte-derived biopesticides

Aquatic macrophyte-derived biopesticides act through a combination of biochemical and physiological mechanisms mediated by allelochemicals such as fatty acids, phenolic compounds, flavonoids, and diterpenes released by aquatic plants [87]. One of the primary mechanisms involves the disruption of microbial or algal cell membranes by saturated and unsaturated fatty acid derivatives (e.g., methyl palmitate), leading to increased permeability, cytoplasmic leakage, and cell lysis. This membrane damage is often coupled with ionic imbalance and oxidative stress, accelerating cell death. Another key mechanism is the inhibition of photosynthesis, particularly Photosystem II, where polyphenolic compounds from *Myriophyllum spicatum* interfere with chlorophyll synthesis and the electron transport chain, causing energy deprivation and suppression of algal proliferation. Additionally, macrophyte-derived metabolites induce oxidative stress through the production of reactive oxygen species, damaging lipids, proteins, and nucleic acids, and enhancing toxicity via synergistic interactions with other allelochemicals.

Macrophytes also release allelopathic exudates – complex mixtures of water-soluble metabolites – into aquatic environments that directly inhibit phytoplankton growth. For example, *Stratiotes aloides* exudates suppress cyanobacterial blooms, while diterpenes from *Potamogeton* spp. function as local antifouling agents by preventing epiphytic algal attachment. Phenolic and tannin-based compounds from *Ceratophyllum demersum* and *Vallisneria natans* further exert broad-spectrum inhibitory effects by targeting enzymatic activity, pigment biosynthesis, and cellular signaling pathways [88].

3. Green synthesis approaches in biopesticide development

Green synthesis is a chemical production approach that builds on the 12 principles of green chemistry, aiming to minimize environmental impacts and enhance sustainability. The concept of green chemistry was first introduced by Anastas and Warner in 1998, and its principles are as follows: Green Chemistry Principle (GCP) 1 – prevention of waste; GCP2 – atom economy (AE); GCP3 – less hazardous chemical synthesis; GCP4 – design of safer chemicals; GCP5 – safer solvents and auxiliaries; GCP6 – energy efficiency; GCP7 – use of renewable feedstocks; GCP8 – reduction of derivatives; GCP9 – catalysis; GCP10 – design for degradation; GCP11 – real-time analysis; and GCP12 – inherently safer processes [89]. In biopesticide production, green synthesis eliminates the need for toxic reducing agents by utilizing the biocatalytic potential of plant extracts and microorganisms. However, the assumption that green-synthesized products are inherently safe is misleading, as differences in nanoparticle size, surface charge, and metal ion release can significantly alter their toxicological behavior and persistence in ecosystems [69, 90]. This approach is critical for significantly reducing the environmental and human health risks associated with conventional chemical synthesis, thus adhering to GCP3, GCP4, and GCP5. Plant-derived metabolites act as natural reducing and stabilizing agents; these metabolites primarily include polyphenols, flavonoids, terpenoids, and reducing sugars. Meanwhile, bacteria, yeasts, molds, and microalgae can interact with metallic salts and convert them into nanoparticles through biological reduction. This biotransformation process not only reduces toxicity but also ensures that the resulting products possess biocompatible, eco-friendly, and sustainable properties. Furthermore, biopolymers such as chitosan, starch, and cellulose are frequently employed as encapsulating or carrier materials. The use of these renewable feedstocks (GCP7) is particularly vital for enabling controlled release of active ingredients and enhancing stability under field conditions [91, 92]. These nanoformulation strategies are especially crucial for new-generation biopesticides such as RNAi-based products, which are often susceptible to rapid degradation by environmental factors (e.g., UV and nucleases). By encapsulating dsRNA using green-synthesized biopolymers, the stability, targeted delivery, and overall field efficacy of the RNAi biopesticides are significantly enhanced. Through the interaction of organisms or plant extracts with metallic salts, nanoparticles can be synthesized biologically, thereby avoiding the use of toxic chemical agents. The products generated through such routes can be applied in both agricultural and environmental settings as eco-friendly nanomaterials. As green products, biopesticides are effective at low concentrations, biodegradable, and harmless to non-target biota. The catalysts used in these processes operate with minimal energy requirements, providing both environmental and economic benefits [93]. Moreover, the physicochemical characteristics of nanoparticles produced via green synthesis – such as surface charge, stability, and coating composition – can significantly influence their biological activity and potential toxicity. Therefore, linking green production routes with comprehensive safety assessments is crucial to ensure the sustainable application of biogenic nanobiopesticides. Beyond agricultural pest management, nanoparticles synthesized through green approaches hold significant promise in diverse application fields. They have been investigated for biomedical uses, energy generation and storage, food safety and preservation, as well as various environmental applications [94, 95].

4. Toxicological profiles of biopesticides

Although biopesticides are often promoted as safer and more environmentally friendly than synthetic pesticides, toxicological assessments in various animal models indicate that their safety cannot be considered absolute. Studies across different taxa have demonstrated that biopesticides can elicit adverse effects in non-target organisms. For instance, in *Danio rerio*, exposure to a natural biopesticide formulated from plant extracts of citronella (*Cymbopogon nardus*), timbo (*Ateleia glazioviana*), annoni grass (*Eragrostis plana*), neem (*Azadirachta indica*), and plane tree (*Platanus occidentalis*) demonstrated mutagenic effects. While conventional herbicides such as glyphosate and atrazine induced significantly higher frequencies of nuclear abnormalities, the natural biopesticide also exhibited genotoxicity compared with controls, albeit to a lower degree than synthetic agents [96].

Similarly, in *Cyprinus carpio*, a 21-day exposure to a Karanjin-based biopesticide derived from *Pongamia pinnata* resulted in significant alterations in biochemical markers, including total proteins, free amino acids, glutamate dehydrogenase, alanine aminotransferase, and aspartate aminotransferase in gill, kidney, liver, and muscle tissues [97]. Developmental toxicity has also been documented. Extracts from *Actiniopteris radiata* were found to be effective against the mosquito vector *Aedes aegypti*; however, concentrations above 250 ppm caused significant mortality in *D. rerio* embryos and induced developmental defects after 72 hours of exposure, including growth retardation, impaired motility, tail deformities, and abnormal ocular development [98].

In contrast, guvermectin, an actinomycete-derived biopesticide (*Streptomyces* sp. NEAU6, endophyte of *Paris polyphylla*), exhibited a relatively low toxicity profile. Acute toxicity assays in *D. rerio* and *Daphnia magna* indicated median lethal concentration/median effective concentration (LC_{50}/EC_{50}) values above 100 mg a.i./L, while its degradation products were even less toxic than the parent compound [99]. Mammalian models have yielded additional insights. In Wistar rats, exposure to both chemical (organophosphate) and biological (bacterial and fungal) pesticides demonstrated hematological disturbances. Notably, fungal-based biopesticides caused an approximately 9% reduction in red blood cell counts, suggesting potential hematotoxicity through disruption of erythrocyte integrity and erythropoiesis. Moreover, all pesticide groups exhibited a 4–5% increase in mean corpuscular hemoglobin, indicating risks of altered erythrocyte morphology and hemoglobin content. Collectively, these findings suggest that although biopesticides may cause milder adverse effects compared with synthetic pesticides, certain groups – particularly fungal-based agents – may exert moderate hematotoxic impacts [100].

Environmental studies have further highlighted risks associated with biopesticides. For example, *Daphnia magna* exposed to the *Bacillus thuringiensis*-based formulation DiPel ES exhibited severe immobilization and mortality in neonates, with EC_{50} values approximately 105-fold lower than those reported by manufacturers. Adults displayed impaired mobility without significant mortality, while biomarker analyses, including body weight, protein, chitobiose, catalase, xenobiotic metabolism, and acetylcholinesterase, indicated disruptions across multiple physiological pathways. Importantly, non-monotonic dose–response relationships (U-shaped or inverted U-shaped) were observed, revealing that even low concentrations may elicit toxic effects and challenging the assumption that Bt-based biopesticides are universally safe [101]. Saponin-rich plant extracts have also raised concerns. Acute toxicity evaluations of saponins extracted from *Quillaja saponaria*, *Chenopodium quinoa*, and *Camellia oleifera* reported hazardous concentration for 5% of species (HC_5) values of 2.91 ± 1.00, 22.9 ± 5.84, and 0.22 ± 0.11 mg/L, respectively. These findings highlight nearly 100-fold differences in toxicity depending on the botanical source. Furthermore, predicted environmental concentrations of these saponins were frequently close to or exceeded water quality standards, suggesting that indiscriminate application may pose significant ecological risks [101]. As with synthetic pesticides, biopesticides may undergo environmental transformations that yield more toxic metabolites than their parent compounds. *Bacillus*-derived proteins, for example, have been detected in aquatic environments adjacent to treated fields. Application frequency and dosage strongly influence the persistence of *Bacillus* crystals, potentially affecting non-target organisms and propagating through food webs. *Bacillus thuringiensis* var.

israelensis crystals can rapidly adsorb onto vegetation and retain high toxicity for up to 22 weeks [102–104].

Despite growing evidence, the toxicological assessment of biopesticides still lacks standardized methodologies comparable to those applied for synthetic pesticides. The absence of harmonized OECD or EFSA testing guidelines for many microbial and botanical agents leads to inconsistent risk characterization. Moreover, the complex mixtures present in crude plant extracts or fermentation products complicate the identification of active toxicants and their degradation products. In some cases, sublethal effects such as oxidative stress, genotoxicity, or endocrine disruption have been documented, yet they are rarely included in regulatory dossiers. Future research should, therefore, integrate mechanistic endpoints and omics-based approaches to elucidate the molecular pathways underlying these responses, enabling evidence-based risk evaluation of biopesticides.

Taken together, these findings indicate that while biopesticides generally present a lower hazard profile compared with synthetic pesticides, their toxicological effects are highly context-dependent. Variables such as plant species, concentration, duration of exposure, and environmental transformations significantly shape their risk potential. Therefore, systematic and mechanistic evaluation of dose–response relationships, chronic sublethal effects, and multi-trophic impacts on non-target species is essential to validate biopesticides as truly sustainable and safe alternatives within IPM frameworks.

5. Global regulatory frameworks for biopesticides

The regulation of biopesticides varies considerably across countries and regions, reflecting national and regional policy priorities. These regulatory frameworks directly affect the commercialization, adoption, and prevalence of biopesticides in agricultural systems. In the United States, the regulation of biopesticides is governed by the Federal Insecticide, Fungicide, and Rodenticide Act and the Federal Food, Drug, and Cosmetic Act. Compared with conventional chemical pesticides, biopesticides are subject to a more streamlined and cost-effective registration process. The Pesticide Registration Improvement Act has significantly reduced both the registration costs and timelines. For instance, the registration of a new biochemical active ingredient for food use costs approximately USD 33,506 with an 18-month review period, whereas a conventional pesticide may cost up to USD 790,737 with a 24-month review period. This regulatory differentiation is intended to facilitate market entry for biopesticides [58]. In the European Union (EU), biopesticides are regulated under Regulation (EC) No 1107/2009 for plant protection products, whereas biostimulants fall under Regulation (EU) 2019/1009. Additionally, the Sustainable Use Directive (2009/128/EC) promotes the adoption of IPM practices that prioritize low-risk alternatives. Strategic policy frameworks such as the European Green Deal and the Farm to Fork Strategy aim to reduce the use and risk of chemical pesticides by 50% by 2030, thereby strengthening the role of biopesticides in sustainable agriculture [105]. In Asia, regulatory approaches vary. In India, biopesticides are regulated under the Insecticides Act (1968), with the registration process managed by the Central Insecticides Board and Registration Committee. Mandatory requirements include DNA fingerprinting and bioefficacy trials conducted at the Indian Council of

Agricultural Research or state agricultural universities. Since 2019, biopesticides can be registered under temporary or permanent categories depending on the completeness of the data submitted [106]. In Japan, biopesticides are regulated under the Agricultural Chemicals Regulation Law. In 1997, the Ministry of Agriculture, Forestry, and Fisheries issued guidelines specific to microbial pesticides, covering bacteria, fungi, viruses, protozoa, and nematodes. A tiered safety assessment system is applied to account for the naturally occurring nature of these organisms [20]. In the Mediterranean region, specific challenges are evident. In Spain, biopesticides are authorized under Regulation (EC) No 1107/2009, consistent with EU legislation. Although biopesticides account for approximately 6% of the market, complex national and EU-level registration procedures restrict their use largely to organic farming systems [107]. In Tunisia, biopesticides are promoted under national IPM policies and through cooperation agreements with the EU, such as the 1995 Barcelona Process. Despite the expansion of plant-based and farmer-prepared formulations, chemical pesticide use remains widespread due to emerging pest pressures such as *Tuta absoluta* [108]. In Turkey, the regulatory framework remains fragmented. The primary legislation is the Plant Protection Products and Pesticide Maximum Residue Limit Regulation (2017), with several active chemical substances recently prohibited (e.g., chlorpropham, oxadiazon in 2022). Although pesticide consumption is relatively high (approximately 52,963 tons per year), the implementation of bans and restrictions has created opportunities for growth in the biopesticide market [109]. Overall, while regulatory frameworks differ across regions, a common trend is the promotion of biopesticides as part of broader sustainability agendas. The cost and time advantages in the United States, the strategic policy targets in the EU, and the tailored guidelines in Asian countries illustrate the diversity of global efforts. These differences, however, underscore the need for greater international harmonization of biopesticide regulations to ensure consistent standards, facilitate global trade, and accelerate the transition toward sustainable agriculture.

6. Limitations and current challenges

Despite the considerable promise of biopesticides as sustainable alternatives to synthetic chemicals, several limitations continue to hinder their broad adoption. Firstly, economic constraints remain critical: The refinement of commercial formulations is expensive, and current manufacturing capacity is inadequate to meet rising global demand. Secondly, technical barriers include the absence of harmonized preparation standards, difficulties in establishing accurate dosage regimes, and the pronounced sensitivity of many products to environmental factors, all of which compromise efficacy and reproducibility [19, 21]. Moreover, compared with synthetic pesticides, most biopesticides exhibit a shorter shelf life and relatively slower modes of action, reducing farmer confidence and market acceptance. From a green chemistry perspective, parameters such as AE and the environmental factor (E-factor) – the ratio of waste generated to product output – provide additional insights into sustainability. While purified biopesticides generally display favorable AE and low E-factor values, several plant-derived preparations generate significant quantities of by-products [110]. Although these residues reduce overall resource efficiency, they can yield secondary ecological benefits when valorized through composting, pathogen suppression, or soil bioremediation strategies [68]. Looking forward, the scaling-up of biopesticides will depend on a multipronged approach: (*i*) reducing production and formulation costs, (*ii*) establishing internationally recognized standards to improve stability and efficacy, and (*iii*) integrating green technologies that minimize waste while maximizing resource efficiency. Such measures will be essential not only for expanding the market penetration of biopesticides but also for aligning their development with global sustainability agendas in agriculture [111].

7. Conclusion

Biopesticides represent essential tools for reducing dependence on synthetic pesticides and are strongly aligned with the United Nations 2030 Sustainable Development Goals (SDGs), particularly SDG 2 (*Zero Hunger*), SDG 3 (*Good Health and Well-being*), SDG 12 (*Responsible Consumption and Production*), and SDG 15 (*Life on Land*). Their development through green synthesis highlights their eco-friendly potential; however, emerging evidence of genotoxicity, developmental toxicity, and unintended ecological impacts indicates that they cannot be regarded as inherently safe. In addition, formulation instability, short shelf life, and limited industrial capacity continue to hinder large-scale adoption and regulatory approval. Overcoming these barriers will require multidisciplinary innovation coupled with supportive policies that incentivize sustainable pest management. Future research should focus on comprehensive toxicological evaluations and the development of harmonized global regulatory standards to ensure that biopesticides are integrated safely and effectively into modern agricultural systems, thereby contributing to the 2030 sustainability agenda.

Nomenclature

AE	Atom economy
Bt	*Bacillus thuringiensis*
dsRNA	Double-stranded RNA
$EC_{50}/LC_{50}/HC_5$	Median effective concentration/median lethal concentration/hazardous concentration for 5% of species
E-factor	Environmental factor
EFSA	European Food Safety Authority
EPA	Environmental Protection Agency
EU	European Union
GABA	Gamma aminobutyric acid
GCP	Green Chemistry Principle
IPM	Integrated pest management
MRL	Maximum Residue Limit
OECD	Organisation for Economic Co-operation and Development
RNAi	RNA interference
SDGs	Sustainable Development Goals
siRNAs	Small interfering RNAs

Conflict of Interest

The authors declare no conflict of interest.

Author details

Derya Kocamaz[1*] and Oguz Merhan[2]

1 Faculty of Arts and Science, Department of Biology, Cukurova University, Adana, Türkiye

2 Faculty of Veterinary, Department of Biochemistry, Kafkas University, Kars, Türkiye

*Address all correspondence to: drykocamaz@gmail.com; dkocamaz@cu.edu.tr

References

[1] Khan BA, Nadeem MA, Nawaz H, Amin MM, Abbasi GH, Nadeem M, et al. Pesticides: Impacts on agriculture productivity, environment, and management strategies. In: Aftab T, editor. Emerging Contaminants and Plants: Interactions, Adaptations and Remediation Technologies. Cham, Switzerland: Springer; 2023. p. 109–134

[2] Ahamad A, Kumar J. Pyrethroid pesticides: An overview on classification, toxicological assessment and monitoring. Journal of Hazardous Materials Advances. 2023;**10**:100284

[3] Refai H, El-Gazar AA, Ragab GM, Hassan DH, Ahmed OS, Hussein RA, et al. Enhanced wound healing potential of Spirulina platensis nanophytosomes: Metabolomic profiling, molecular networking, and modulation of HMGB-1 in an excisional wound rat model. Marine Drugs. 2021;**21**(3):149

[4] Umapathi R, Ghoreishian SM, Sonwal S, Rani GM, Huh YS. Portable electrochemical sensing methodologies for on-site detection of pesticide residues in fruits and vegetables. Coordination Chemistry Reviews. 2022;**453**:214305

[5] Kitchen LW, Lawrence KL, Coleman RE. The role of the United States military in the development of vector control products, including insect repellents, insecticides, and bed nets. Journal of Vector Ecology. 2009;**34** (1):50–61

[6] Tudi M, Daniel Ruan H, Wang L, Lyu J, Sadler R, Connell D, et al. Agriculture development, pesticide application and its impact on the environment. International Journal of Environmental Research and Public Health. 2021;**18**(3):1112

[7] Cech R, Zaller JG, Lyssimachou A, Clausing P, Hertoge K, Linhart C. Pesticide drift mitigation measures appear to reduce contamination of non-agricultural areas, but hazards to humans and the environment remain. Science of the Total Environment. 2023;**854**:158814

[8] Ahmed Z, Crush J, Owuor S, Onyango EO. Disparities and determinants of Somali refugee food security in Nairobi, Kenya. Global Food Security. 2024;**43**:100808

[9] Bakır A, Erdag D, Elis Yildiz S, Karadag Sarı E, Asker H, Sözmen M. Immunohistochemical examination on the effects of malathion and *Onosma nigricaule* (Boraginaceae) on the catalase (CAT) and superoxide dismutase-2 (Mn-SOD) in renal tissues of mice. Ankara University Veterinary Faculty Journal. 2017;**64**:125–130

[10] Mostafalou S, Abdollahi M. Pesticides and human chronic diseases: Evidences, mechanisms, and perspectives. Toxicology and Applied Pharmacology. 2013;**268**(2):157–177

[11] Zhang N, Zhu L, Zhang R, Zhang C, Cheng J, Tao L, et al. Evaluation of toxicological effects of organophosphorus pesticide metabolites on human HepG2 cells. Environmental Toxicology and Pharmacology. 2021;**88**:103741

[12] Sinaci C, Celik A, Yetkin D, Cevik S, Guler G. Sulfoxaflor insecticide exhibits cytotoxic or genotoxic and

apoptotic potential via oxidative stress-associated DNA damage in human blood lymphocytes cell cultures. Drug and Chemical Toxicology. 2023;**46**(5):972–983

[13] Cavalier H, Trasande L, Porta M. Exposures to pesticides and risk of cancer: Evaluation of recent epidemiological evidence in humans and paths forward. International Journal of Cancer. 2023;**152**(5):879–912

[14] Scorza FA, Beltramim L, Bombardi LM. Pesticide exposure and human health: Toxic legacy. Clinics (Sao Paulo). 2023;**78**:100249

[15] Aktac T, Bakar E, Guner U. The effects of short-term exposure to cadmium and copper on sialic acid in carp (*Cyprinus carpio*) tissues. Fresenius Environmental Bulletin. 2010;**19**(3):432–437

[16] Borges S, Alkassab AT, Collison E, Hinarejos S, Jones B, McVey E, et al. Overview of the testing and assessment of effects of microbial pesticides on bees: Strengths, challenges and perspectives. Apidologie. 2021; **52**:1256–1277

[17] Glare TR, Gwynn RL, Moran-Diez ME. Development of biopesticides and future opportunities. In: Moran-Diez ME, Glare TR, editors. Microbial-based Biopesticides: Methods and Protocols. Vol. **1477**: New York: Humana Press; 2016. p. 211–221

[18] Karakas M. Some Important Herbal Insecticides. Turkish Journal of Scientific Reviews. 2018;**11**:32–37

[19] Mishra J, Dutta V, Arora N. Biopesticides in India: Technology and sustainability linkages. 3 Biotech. 2020;**10**(5):210

[20] Yokoyama T. Development and use of biofertilizers in Japan: current situations and future directions. Asia-Pacific Biofertilizers and Biopesticides Information Platform. 2023. DOI: 10.56669/TGJS1978

[21] Stevenson PC, Isman MB, Belmain SR. Pesticidal plants in Africa: A global vision of new biological control products from local uses. Industrial Crops and Products. 2017;**110**:2–9

[22] Freimoser FM, Rueda-Mejia MP, Tilocca B, Migheli Q. Biocontrol yeasts: Mechanisms and applications. World Journal of Microbiology and Biotechnology. 2019;**35**(10):154

[23] Palermo D, Giunti G, Laudani F, Palmeri V, Campolo O. Essential oil-based nano-biopesticides: Formulation and bioactivity against the confused flour beetle Tribolium confusum. Sustainability. 2021;**13**(17):9746

[24] Jadhav KB, Nagarkar JM. Development of karanj oil and castor oil nanoemulsions and its encapsulation into beads for controlled release larvicidal activity. Journal of Advanced Scientific Research. 2022;**13**(02):93–100

[25] Pascual-Villalobos MJ, Guirao P, Díaz-Baños FG, Cantó-Tejero M, Villora G. Oil in water nanoemulsion formulations of botanical active substances. In: Koul O, editor. Nano-biopesticides Today and Future Perspectives. London: Academic Press; 2019. p. 223–247

[26] Hazafa A, Murad M, Masood MU, Bilal S, Khan MN, Farooq Q, et al. Nano-

biopesticides as an emerging technology for pest. In: El-Shafie AF, editor. Insecticides: Impact and Benefits of Its Use for Humanity. IntechOpen: London; 2022. p. 407–425

[27] Essiedu JA, Adepoju FO, Ivantsova MN. Benefits and limitations in using biopesticides: A review. In: Proceedings of the VII International Conference of Young Researchers – Physics, Technology, Innovations. Yekaterinburg, Russia. 2020; Melville, USA: AIP Publishing; 2020:2313:080002

[28] Wilson K, Grzywacz D, Curcic I, Scoates F, Harper K, Rice A, et al. A novel formulation technology for baculoviruses protects biopesticide from degradation by ultraviolet radiation. Scientific Reports. 2020;**10** (1):13301

[29] Bashan Y, Hernandez J-P, Leyva LA, Bacilio M. Alginate microbeads as inoculant carriers for plant growth-promoting bacteria. Biology and Fertility of Soils. 2002;**35** (5):359–368

[30] Anastas ND, Warner JC. The incorporation of hazard reduction as a chemical design criterion in green chemistry. Chemical Health & Safety. 2005;**12**(2):9–13

[31] Ayilara MS, Adeleke BS, Akinola SA, Fayose CA, Adeyemi UT, Gbadegesin LA, et al. Biopesticides as a promising alternative to synthetic pesticides: A case for microbial pesticides, phytopesticides, and nanobiopesticides. Frontiers in Microbiology. 2023;**14**:1040901

[32] Sundararajan B, Kumari BDR. Novel synthesis of gold nanoparticles using *Artemisia vulgaris* L. leaf extract and their efficacy of larvicidal activity against dengue fever vector Aedes aegypti L. Journal of Trace Elements in Medicine and Biology. 2017;**43**:187–196

[33] Musawa BB, Muhammad MS, Onyema R. Modeling and distribution of neem (Azadirachta indica a. juss) in Nigeria Katsina Journal of Natural and Applied Sciences. **10**(2):2141–0755

[34] Lallawmawma H, Sathishkumar G, Sarathbabu S, Ghatak S, Sivaramakrishnan S, Gurusubramanian G, et al. Synthesis of silver and gold nanoparticles using Jasminum nervosum leaf extract and its larvicidal activity against filarial and arboviral vector Culex quinquefasciatus say (Diptera: Culicidae). Environmental Science and Pollution Research 2015;**22**:17753–17768

[35] Vinci G, Savastano M, Restuccia D, Ruggeri M. Nanobiopesticides: Sustainability aspects and safety concerns. Environments. 2025;**12**(3):74

[36] Cherif A, Mansour R, Sun C, Grissa-Lebdi K. Lethal effects of nano and commercial formulations of abamectin on Tuta absoluta (Meyrick) and its mirid predators Macrolophus pygmaeus and Nesidiocoris tenuis. International Journal of Tropical Insect Science. 2022;**42**(3):2183–2193

[37] Pan X, Guo X, Zhai T, Zhang D, Rao W, Cao F, et al. Nanobiopesticides in sustainable agriculture: Developments, challenges, and perspectives. Environmental Science: Nano 2023;**10**:41–61

[38] De Neef E, Velásquez-Zapata V, Gordon ERL, Narva K, Mc Cahon P, Mezin L, et al. A bioinformatic ecological risk assessment framework for externally applied double-stranded RNA-based biopesticides. Integrated

Environmental Assessment and Management. 2025;**33**:1–16

[39] Ghosson H, Salvia M-V, Bertrand C. Development of omics tools for the assessments of the environmental fate and impact of biocontrol agents. In: Priget-Combaret C, Dumas B, editors. Biocontrol of Plant Disease: Recent Advances and Prospects in Plant Protection. New York: Wiley; 2022. pp. 47–73

[40] OECD. Report of the 12th Expert Group on Biopesticides Seminar on Problem Formulation for the risk assessment of biopesticides, OECD Series on Pesticides and Biocides, OECD Publishing, Paris. 2024

[41] Kumar J, Ramlal A, Mallick D, Mishra V. An overview of some biopesticides and their importance in plant protection for commercial acceptance. Plants. 2021;**10**(6):1185

[42] Ali AM, Mohamed DS, Shaurub E, Elsayed AM. Antifeedant activity and some biochemical effects of garlic and lemon essential oils on Spodoptera littoralis (Boisduval) (Lepidoptera: Noctuidae). Journal of Entomology and Zoology Studies. 2017;**5**(3):1476–1482

[43] Okwute SK. Plants as potential sources of pesticidal agents: A review. In: Soundararajan RP, editor. Pesticides: Advances in Chemical and Botanical Pesticides. London: IntechOpen. 2012. pp. 207–320

[44] Souto AL, Sylvestre M, Tölke ED, Tavares JF, Barbosa-Filho JM, Cebrián-Torrejón G. Plant-Derived Pesticides as an Alternative to Pest Management and Sustainable Agricultural Production: Prospects, Applications and Challenges. Molecules. 2021;**26**(16):4835

[45] Gakuubi MM, Maina AW, Wagacha JM. Antifungal activity of essential oil of Eucalyptus camaldulensis Dehnh. against selected Fusarium spp. International Journal of Microbiology. 2017;**2017**:8761610

[46] Ahmad W, Singh S, Kumar S. Phytochemical screening and antimicrobial study of Euphorbia hirta extracts. Journal of Medicinal Plants Studies. 2017;**5**(2):183–186

[47] Sawangproh W, Paejaroen P, Afifah L, Phaenark C. Microbial pesticides: A bibliometric analysis of global research trends (1973–2024). Egyptian Journal of Biological Pest Control. 2025;**35**(1):2

[48] Lin F, Mao Y, Zhao F, Idris AL, Liu Q, Zou S, et al. Towards sustainable green adjuvants for microbial pesticides: Recent progress, upcoming challenges, and future perspectives. Microorganisms 2023;**11**:364

[49] Tiwari AK. Insect pests in agriculture identifying and overcoming challenges through IPM. Archives of Current Research International. 2024;**24**(3):124–130

[50] Ray PD, Huang B-W, Tsuji Y. Reactive oxygen species (ROS) homeostasis and redox regulation in cellular signaling. Cellular Signalling. 2012;**24**(5):981–990

[51] Sarkar A, Pramanik K, Mitra S, Soren T, Maiti TK. Enhancement of growth and salt tolerance of rice seedlings by ACC deaminase-producing Burkholderia sp. MTCC 12259. Journal of Plant Physiology. 2018;**231**:434–442

[52] Weyens N, Van Der Lelie D, Taghavi S, Newman L, Vangronsveld J. Exploiting plant–microbe partnerships

to improve biomass production and remediation. Trends in Biotechnology. 2009;**27**(10):591–598

[53] Gomiero T. Food quality assessment in organic vs. conventional agricultural produce: Findings and issues. Applied Soil Ecology. 2018;**123**:714–728

[54] Ranglová K, Lakatos GE, Manoel JAC, Grivalský T, Estrella FS, Fernández FGA, et al. Growth, biostimulant and biopesticide activity of the MACC-1 Chlorella strain cultivated outdoors in inorganic medium and wastewater. Algal Research. 2021;**53**:102136

[55] Machado S, Pereira R, Sousa RMO. Nanobiopesticides: Are they the future of phytosanitary treatments in modern agriculture? Science of the Total Environment. 2023;**896**:166401

[56] Fu L, Wang Z, Dhankher OP, Xing B, Foyer C. Nanotechnology as a new sustainable approach for controlling crop diseases and increasing agricultural production. Journal of Experimental Botany. 2020;**71**(2): 507–519

[57] Su Y, Ashworth V, Kim C, Adeleye AS, Rolshausen P, Roper C, et al. Delivery, uptake, fate, and transport of engineered nanoparticles in plants: A critical review and data analysis. Environmental Science: Nano 2019;**6**:2311–2331

[58] Farooq MA, Hannan F, Islam F, Ayyaz A, Zhang N, Chen W, et al. The potential of nanomaterials for sustainable modern agriculture: Present findings and future perspectives. Environmental Science: Nano 2022;**9**:1926–1951

[59] Ermini ML, Voliani V. Antimicrobial nano-agents: The copper age. ACS Nano. 2021;**15**(4):6008–6029

[60] Granetto M, Serpella L, Fogliatto S, Re L, Bianco C, Vidotto F, et al. Natural clay and biopolymer-based nanopesticides to control the environmental spread of a soluble herbicide. Science of the Total Environment. 2022;**806**:151199

[61] Hamilton AJ, Baulcombe DC. A species of small antisense RNA in posttranscriptional gene silencing in plants. Science. 1999;**286**(5441): 950–952

[62] Zamore PD, Tuschl T, Sharp PA, Bartel DP. RNAi: Double-stranded RNA directs the ATP-dependent cleavage of mRNA at 21 to 23 nucleotide intervals. Cell. 2000;**101**(1):25–33

[63] Dalmay T, Hamilton A, Rudd S, Angell S, Baulcombe DC. An RNA-dependent RNA polymerase gene in Arabidopsis is required for posttranscriptional gene silencing mediated by a transgene but not by a virus. Cell. 2000;**101** (5):543–553

[64] Mao Y-B, Cai W-J, Wang J-W, Hong G-J, Tao X-Y, Wang L-J, et al. Silencing a cotton bollworm P450 monooxygenase gene by plant-mediated RNAi impairs larval tolerance of gossypol. Nature Biotechnology 2007;**25**:1307–1313

[65] Baumberger N, Baulcombe DC. Arabidopsis ARGONAUTE1 is an RNA Slicer that selectively recruits microRNAs and short interfering RNAs. In: Proceedings of the National Academy of Sciences. 2005;**102** (33):11928–11933

[66] Head GP, Carroll MW, Evans SP, Rule DM, Willse AR, Clark TL, et al. Evaluation of SmartStax and SmartStax PRO maize against western corn rootworm and northern corn rootworm: Efficacy and resistance management. Pest Management Science 2017;**73**: 1883–1899

[67] Baum JA, Bogaert T, Clinton W, Heck GR, Feldmann P, Ilagan O, et al. Control of coleopteran insect pests through RNA interference. Nature Biotechnology 2007;**25**:1322–1326

[68] Taning CNT, Gui S, De Schutter K, Jahani M, Castellanos NL, Christiaens O, et al. A sequence complementarity-based approach for evaluating off-target transcript knockdown in Bombus terrestris, following ingestion of pest-specific dsRNA. Journal of Pest Science 2021;**94**: 487–503

[69] Sundh I, Goettel MS. Regulating biocontrol agents: A historical perspective and a critical examination comparing microbial and macrobial agents. BioControl. 2013;**58**(5):575–593

[70] Damalas CA, Koutroubas SD. Current status and recent developments in biopesticide use. Agriculture. 2018;**8** (1):13

[71] Kumar S, Singh A. Biopesticides: Present status and the future prospects. Journal of Fertilizers & Pesticides. 2015;**6**(2):1000e129

[72] Fu Y, Bhadha JH, Rott P, Beuzelin JM, Kanissery R. Investigating the use of aquatic weeds as biopesticides towards promoting sustainable agriculture. PloS One. 2020;**15**(8): e0237258

[73] Benvenuti S, Cioni PL, Flamini G, Pardossi A. Weeds for weed control: Asteraceae essential oils as natural herbicides. Weed Research. 2017;**57** (5):342–353

[74] Ismail A, Lamia H, Mohsen H, Samia G, Bassem J. Chemical composition, bio-herbicidal and antifungal activities of essential oils isolated from Tunisian common cypress (Cupressus sempervirens L.). Journal of Medicinal Plants Research. 2013;**7** (16):1070–1080

[75] Ismail A, Habiba K, Yassine M, Mohsen H, Bassem J, Lamia H. Essential oils of Tunisian Pinus radiata D. Don, chemical composition and study of their herbicidal activity. Vietnam Journal of Chemistry. 2021;**59**(2):247–252

[76] Pouresmaeil M, Nojadeh MS, Movafeghi A, Maggi F. Exploring the bio-control efficacy of Artemisia fragrans essential oil on the perennial weed Convolvulus arvensis: Inhibitory effects on the photosynthetic machinery and induction of oxidative stress. Industrial Crops and Products. 2020;**155**:112785

[77] Sarma R, Khanikor B, Mahanta S. Essential oil from Citrus grandis (Sapindales: Rutaceae) as insecticide against Aedes aegypti (L)(Diptera: Culicidae). International Journal of Mosquito Research. 2017;**4** (3):88–92

[78] Kostyukovsky M, Rafaeli A, Gileadi C, Demchenko N, Shaaya E. Activation of octopaminergic receptors by essential oil constituents isolated from aromatic plants: Possible mode of action against insect pests. Pest Management Science. 2002;**58** (11):1101–1106

[79] Gaire S, Scharf ME, Gondhalekar AD. Toxicity and neurophysiological impacts of plant essential oil components on bed bugs (Cimicidae: Hemiptera). Scientific Reports. 2019;**9**(1):3961

[80] Rasooli I, Rezaei MB, Allameh A. Growth inhibition and morphological alterations of Aspergillus Niger by essential oils from Thymus eriocalyx and Thymus x-porlock. Food Control. 2006;**17**(5):359–364

[81] Kumar S. Biopesticides: A Need for Food and Environmental Safety. Journal of Biofertilizers & Biopesticides. 2012;**3**:e107

[82] Copping LG, Menn JJ. Biopesticides: A review of their action, applications and efficacy. Pest Management Science. 2000;**56**(8):651–676

[83] Grillo R, Fraceto LF, Amorim MJB, Scott-Fordsmand JJ, Schoonjans R, Chaudhry Q. Ecotoxicological and regulatory aspects of environmental sustainability of nanopesticides. Journal of Hazardous Materials. 2021;**404**:12414

[84] Machado S, Pereira R, Sousa RMOF. Nanobiopesticides: Are they the future of phytosanitary treatments in modern agriculture? Science of the Total Environment. 2023;**896**:166401

[85] Huvenne H, Smagghe G. Mechanisms of dsRNA uptake in insects and potential of RNAi for pest control: A review. Journal of Insect Physiology. 2010;**56**(3):227–235

[86] Whyard S, Singh AD, Wong S. Ingested double-stranded RNAs can act as species-specific insecticides. Insect Biochemistry and Molecular Biology. 2009;**39**(11):824–832

[87] Macías FA, Molinillo JMG, Varela RM, Galindo JCG. Allelopathy—a natural alternative for weed control. Pest Management Science. 2007;**63**(4):327–348

[88] Fu Y, Bhadha JH, Rott P, Beuzelin JM, Kanissery R. Investigating the use of aquatic weeds as biopesticides towards promoting sustainable agriculture. Plos One. 2020;**15**(8): e0237258

[89] Fenibo EO, Ijoma GN, Nurmahomed W, Matambo T. The potential and green chemistry attributes of biopesticides for sustainable agriculture. Sustainability. 2022;**14**(21):14417

[90] Yatagai M, Nishimoto M, Hori K, Ohira T, Shibata A. Termiticidal activity of wood vinegar, its components and their homologues. Journal of Wood Science. 2002;**48**(4):338–342

[91] Zhao X, Hou N, Wan C, Zhang L, Liu X. Gold nanoparticles synthesis mediated by fungus isolated from aerobic granular sludge: Process and mechanisms. Heliyon. 2024;**10**(6): e28281

[92] Ekinci A, Şahin Ö, Kutluay S, Horoz S, Canpolat G, Çokyaşa M, et al. Designing copper-doped zinc oxide nanoparticle by tobacco stem extract-mediated green synthesis for solar cell efficiency and photocatalytic degradation of methylene blue. International Journal of Phytoremediation 2024;**26**:2183–2193

[93] Tadesse Mawcha K, Malinga L, Muir D, Ge J, Ndolo D. Recent advances in biopesticide research and development with a focus on microbials. F1000Research. 2025;**13**:1071

[94] Hezakiel HE, Thampi M, Rebello S, Sheikhmoideen JM. Biopesticides: A green approach towards agricultural pests. Applied Biochemistry and Biotechnology. 2024;**196**(8):5533–5562

[95] Altammar KA. A review on nanoparticles: Characteristics, synthesis, applications, and challenges. Frontiers in Microbiology. 2023;**14**:1155622

[96] Lechinovski L, Bados M, Rosa J, Moda DB, Bueno Krawczyk ACDD. Ecotoxicological effects of conventional herbicides and a natural herbicide on freshwater fish (Danio rerio). Journal of Environmental Science and Health, Part B. 2022;**57**(10):812–820

[97] Tasneem S, Yasmeen R. Biochemical alterations in total proteins and related enzymes in tissues of *Cyprinus carpio* (L.) during sublethal exposure to karanjin based biopesticide Derisom. Indian Journal of Experimental Biology. 2021;**59**:125–131

[98] Kamaraj C, Deepak P, Balasubramani G, Karthi S, Arul D, Aiswarya D, et al. Target and non-target toxicity of fern extracts against mosquito vectors and beneficial aquatic organisms. Ecotoxicology and Environmental Safety. 2018;**161**: 221–230

[99] Shi Y, An X, Zhang B, Pan X, Wu X, Xu J, et al. Hydrolysis, photolysis, and biotoxicity assessment of a novel biopesticide, guvermectin. Journal of Agricultural and Food Chemistry 2022;**70**:16117–16125

[100] Faokunla O, Muhammad SS, Owoeye FD, Kelly BA, Amoo OT, Ajayi BM, Abel IG, Adejoh VU, Adeoye TM, Ojo DB, Jimoh T, Luke DP, Ajayi AM, Oladimeji FB, Alfred TO. Assessment of haemato-toxicity induced by chemical and biological pesticides in grain-fed male Wistar rats. Confluence Journal of Pure and Applied Sciences. 2025;**4**:105–121

[101] Jiang X, Hansen HCB, Strobel BW, Cedergreen N. What is the aquatic toxicity of saponin-rich plant extracts used as biopesticides? Environmental Pollution. 2018;**236**:416–424

[102] Chen S, Dzewaltowski DA, Rosenkranz RR, Lanningham-Foster L, Vazou S, Gentile DA, et al. Feasibility study of the SWITCH implementation process for enhancing school wellness. BMC Public Health 2018;**18**:1119

[103] Dhaouadi S, Jeni RE, Kraiem H, Ayyildiz G, Filik-Iscen C, Yurtkuran-Ceterez Z, et al. Effects of new Btk-based formulations BLB1 and lip on aquatic non-target organisms. Biology 2024;**13**:824

[104] Boisvert M, Boisvert J. Effects of Bacillus thuringiensis var. israelensis on target and nontarget organisms: A review of laboratory and field experiments. Biocontrol Science and Technology. 2000;**10**(5):517–561

[105] Birch EAN, Begg GS, Squire GR. How agro-ecological research helps to address food security issues under new IPM and pesticide reduction policies for global crop production systems. Journal of Experimental Botany. 2011;**62**: 3251–3261

[106] Chakraborty N, Mitra R, Pal S, Ganguly R, Acharya K, Minkina T, et al. Biopesticide consumption in India: insights into the current trends. Agriculture 2023;**13**:557

[107] Khursheed A, Rather MA, Jain V, Wani AR, Rasool S, Nazir R, et al. Plant based natural products as potential ecofriendly and safer biopesticides: A comprehensive overview of their advantages over conventional pesticides, limitations and regulatory aspects. Microbial Pathogenesis. 2022;**173**:105854

[108] Fusar Poli E, Fontefrancesco MF. Trends in the implementation of biopesticides in the Euro-Mediterranean region: A narrative literary review. Sustainable Earth Reviews. 2024;**7**(1):14

[109] Hanedar AT, Girgin E. Pesticide Management within the Context of Green Deal and Türkiye. ITU Çevre İklim ve Sürdürülebilirlik. 2023;**24** (2):87–96

[110] Jessop PG, Trakhtenberg S, Warner J. The Twelve Principles of Green Chemistry, Innovations in Industrial and Engineering Chemistry. In: Flank WH, Abraham MA, Matthews MA, editors. Washington: ACS Publications; 2009. p. 401–436

[111] Fenibo EO, Ijoma GN, Matambo T. Biopesticides in sustainable agriculture: A critical sustainable development driver governed by green chemistry principles. Frontiers in Sustainable Food Systems. 2021;**5**:619058

Chapter 6

RNA Interference Technology: From Molecular Mechanisms to Applications as Green Pesticides

Yiying Zhao, Xiaojun Wang, Kedi Zhao, Qiancheng Wei and Feng Liu

Abstract

Addressing the issues of pesticide resistance, environmental pollution, and health risks associated with conventional chemical pesticides necessitates the development of efficient and environmentally benign pest control strategies. RNA interference (RNAi) technology, particularly double-stranded RNA (dsRNA)-based biopesticides, offers a promising alternative due to its high species specificity and biodegradable nature, potentially balancing effective pest control with environmental and health safety. This chapter comprehensively reviews the mechanism of action, advances in delivery strategies (e.g., transgenic plants, foliar sprays, and nano-carriers), and current applications of dsRNA pesticides. It critically analyzes the stability challenges dsRNA encounters in the environment and within insect bodies, alongside corresponding mitigation strategies. Furthermore, the chapter examines potential ecological risks, regulatory hurdles, and future perspectives, underscoring the need for integrated technological innovation and thorough assessment to establish RNAi as a cornerstone of sustainable agricultural pest management.

Keywords: RNA interference, double-stranded RNA, green pesticides, nanocarriers, environmental safety

1. Introduction

Agricultural pests account for up to 40%–50% of global crop yield losses annually. While the extensive use of chemical pesticides has partially mitigated pest damage, it has also led to issues such as pesticide resistance, environmental pollution, and health risks [1]. In this context, there is an urgent need to develop green, efficient, and highly specific bio-pesticides. RNA interference (RNAi) technology, particularly nucleic acid-based pesticides using double-stranded RNA (dsRNA), is regarded as a potential alternative to conventional chemical pesticides due to its high specificity, perceived environmental friendliness, and biodegradable nature [2, 3], though its practical efficacy and safety require comprehensive evaluation. This chapter provides a systematic review of the mechanism of action, delivery strategies, environmental stability, current applications, and challenges of RNAi-based bio-pesticides in pest control, along with prospects for future development [4, 5].

2. Mechanism of action and advantages of dsRNA bio-pesticides

The RNAi is a relatively conserved and widespread post-transcriptional gene regulatory mechanism in eukaryotes. It operates by specifically degrading target mRNA, leading to gene silencing or suppression of gene expression, and is highly conserved across species of different evolutionary statuses. Specifically, dsRNA is randomly cleaved by Dicer-like proteins into small RNAs (siRNAs or miRNAs) with lengths of 21–24 nucleotides. The siRNAs then bind to Argonaute proteins (AGOs) to form the RNA-induced silencing complex (RISC), which complements the target RNA strand, inducing mRNA degradation or translational inhibition [6–8]. By targeting essential genes of pests through RNAi, efficient gene silencing can be achieved, thereby effectively controlling the occurrence of diseases and insect pests [9]. In the field of entomology, RNAi technology not only provides a convenient tool for functional genomic studies of insects but also offers a novel approach for controlling agricultural pests [10].

Although the RNAi machinery is relatively conserved in eukaryotes, its efficiency varies dramatically across insect orders, primarily due to two key factors: dsRNA uptake pathways and the integrity of intracellular systems. In insects with high RNAi efficiency, such as Coleoptera and Hemiptera, a systemic RNAi response mediated by transmembrane channel proteins (e.g., Sid-1 homologs) allows efficient intercellular transmission of dsRNA. However, in many RNAi-recalcitrant pests like Lepidoptera, such efficient systemic spreading mechanisms are often absent, and cells primarily rely on less efficient endocytosis for dsRNA uptake. More critically, the hemolymph and midgut of lepidopteran insects contain abundant and active dsRNA-specific nucleases (dsRNases) that rapidly degrade exogenous dsRNA before it reaches target cells. Furthermore, the alkaline gut environment, complex microbiota, and the peritrophic matrix constitute additional physical and biochemical barriers, further compromising dsRNA stability and cellular uptake. Therefore, recognizing this significant interspecific variation is a prerequisite for designing and developing effective RNAi-based pesticides.

Compared to traditional chemical pesticides, dsRNA-based biopesticides offer several theoretical advantages: theoretically high species specificity, acting only on target pests while being safe for non-target organisms (such as bees and natural enemies); perceived environmental friendliness, as they are readily degradable under natural conditions without residual risks; and a short development cycle – once the target gene is identified, the corresponding dsRNA can be quickly designed and synthesized without complex chemical processes. Furthermore, multi-target design can delay the development of pest resistance [11]. In 2016 and 2017, Canada and the United States approved the first commercial RNAi-based insecticide, SmartStax® PRO, which uses transgenic corn to express dsRNA targeting the Dvsnf7 gene of the western corn rootworm (Diabrotica virgifera), marking a major breakthrough in the application of RNAi technology for pest control [12]. Furthermore, in the EU, sprayable dsRNA products targeting the Colorado Potato Beetle (e.g., the proof-of-concept for BioDirect) are undergoing advanced field trials and comprehensive environmental risk assessments. Concurrently, research institutions and enterprises in China have made notable experimental progress in developing RNAi rice varieties against pests like the brown planthopper and in advancing nanocarrier delivery systems, demonstrating the global potential and adaptability of this technology.

However, alongside its significant application potential, the potential ecological risks of RNAi technology, particularly dsRNA-based biopesticides, warrant careful consideration and assessment. A primary concern is off-target effects. Due to potential partial sequence homology, dsRNA designed to silence specific pest genes may inadvertently affect non-target organisms, such as beneficial insects (e.g., pollinating bees, parasitic wasps) or soil arthropods, potentially disrupting the balance of the agricultural ecosystem. Secondly, the environmental fate and persistence of exogenously applied dsRNA are not fully understood. These dsRNA molecules may persist in soil or water systems and be taken up by soil microbiota or aquatic organisms, leading to unforeseen impacts on soil ecological functions and aquatic ecosystems. Therefore, before promoting RNAi technology as a "green" solution, it is imperative to systematically evaluate its long-term effects on the entire agricultural ecosystem through rigorous laboratory studies and field monitoring to ensure its environmental safety.

Despite these advantages, the application of nucleic acid pesticides in transgenic crops still faces multiple limitations, including low public acceptance, complex regulatory approval processes, and lengthy research and development cycles [13]. To circumvent these constraints, researchers have developed non-transgenic delivery strategies, such as sprayable exogenous dsRNA to induce gene silencing in pests. In recent years, the emergence of chloroplast transformation technology has further enhanced the stability of dsRNA in plants by avoiding recognition and degradation by the plant's own Dicer enzymes, thereby improving RNAi efficiency [14]. However, this technology still has limitations: its applicability is restricted to certain species, chloroplast transformation remains challenging in major crops such as rice, wheat, and corn, market acceptance is relatively low, and potential environmental risks require stringent assessment [15]. Moreover, products derived from transgenic crops generally face low public acceptance in most countries, and their potential environmental impacts must undergo rigorous risk evaluation [16].

3. Delivery strategies and technological advances of dsRNA

3.1 Transgenic plant-mediated RNAi

Early studies utilized transgenic plants to express dsRNA targeting pest genes. For example, cotton engineered with the P450 gene effectively suppressed larval feeding in Helicoverpa armigera [17]. However, the widespread adoption of such crops has been limited due to low public acceptance, lengthy regulatory approval processes, and high technical barriers [18].

3.2 Foliar spray and SIGS technology

Spray-induced gene silencing (SIGS) involves the topical application of dsRNA, which is absorbed and systemically transported within the plant, enabling the control of hidden or sheltered pests. Studies have detected sprayed dsRNA in vascular tissues, leaves, and roots, demonstrating its efficacy in silencing target pest genes [19]. Bayer's BioDirect product, currently under review by the U.S. Environmental Protection Agency (EPA), represents a significant step toward the commercialization of sprayable nucleic acid pesticides [20].

3.3 Nanocarrier-mediated delivery systems

Compared to conventional delivery methods, nanocarrier-mediated RNAi systems offer advantages such as high efficiency, stability, low dosage requirements, and sustained release. The delivery process primarily involves four steps [21]: nanocarrier-nucleic acid complexation, cellular uptake, endosomal escape, and dissociation of the nanocomplex, followed by nucleic acid release. Specifically, nanocarriers bind to the phosphate groups of dsRNA via electrostatic interactions to form nucleic acid-nanoparticle complexes. These complexes are internalized via endocytosis, forming endocytic vesicles that traffic along microtubules and fuse with early endosomes. To avoid degradation in lysosomes, the complexes must escape from the endosomes into the cytoplasm. Within the anion-rich intracellular environment, the nanocarriers dissociate, releasing dsRNA or siRNA, which then engages the insect RNAi machinery to degrade target mRNA. This approach significantly enhances nucleic acid protection and delivery efficiency compared to traditional strategies [22]. Li et al. [23] synthesized a star-shaped polycation (SPc) using commercially available and low-cost materials, which exhibited low cytotoxicity, high gene delivery efficiency, and excellent stability. SPc-dsRNA complexes effectively reduced target gene expression and inhibited growth in Agrotis ipsilon larvae. In another study, transgenic tobacco plants expressing dsRNAs and hairpin RNAs (hpRNAs) targeting four essential genes (ACT, TUB, VAT, and SNF) of Frankliniella occidentalis induced stronger RNAi responses and higher insect mortality compared to nuclear-transformed plants [24]. Graphene oxide (GO) as a nanocarrier, when combined with pyrethroid insecticides, significantly enhanced acaricidal activity against Tetranychus urticae in both greenhouse and field trials, with synergy ratios (SR) ranging from 2.67 to 3.55 [25]. The GO–pesticide complex also demonstrated improved sustained release properties, enhanced UV resistance, extended half-life, and greater physical stability. Layered double hydroxide (LDH) nanomaterials effectively protect biomolecules such as DNA and RNA on leaf surfaces and enable controlled release for prolonged activity [26]. A single-walled carbon nanotube (SWCNT)-based delivery system achieved up to 95% gene silencing efficiency by efficiently delivering siRNA into plant cells; the silencing effect commenced within one day and persisted for approximately seven days [27]. Liu and colleagues [28] further demonstrated that single-stranded DNA–SWCNT complexes can penetrate plant cell walls. SWCNT-based systems can deliver nucleic acids such as plasmid DNA (pDNA) not only into plant cells but also into specific organelles, including chloroplasts and mitochondria, offering new avenues for functional genomics and crop protection [29].

Beyond the described nanocarriers, viral vector systems represent another highly efficient strategy for the *in vivo* delivery of dsRNA. Among them, baculoviruses and plant viruses (e.g., Tobacco Mosaic Virus, TMV) are extensively engineered as biotechnology platforms. Baculoviruses possess natural tropism for insects, and their capsid proteins can be modified to enhance targeting specificity toward midgut cells of particular pests. Through genetic engineering, the baculovirus genome can be designed to integrate and express dsRNA or hpRNA targeting essential pest genes. Upon infection, the virus replicates extensively within host cells, enabling continuous and robust production of RNAi triggers, thereby inducing potent and sustained gene silencing. In contrast, plant virus vectors, such as virus-like particles (VLPs), offer the advantage of systemic mobility within the plant, acting as protective "nanocontainers" to deliver pre-loaded dsRNA throughout the entire plant. For instance, the coat protein of a modified Cowpea chlorotic mottle virus (CPMV) can self-assemble into hollow structures encapsulating dsRNA, and its surface can be functionalized to promote endocytosis by plant cells. Although viral vectors are exceptionally efficient, their safety assessment – including host range and potential risks of genetic recombination – and the cost of large-scale production are critical considerations that require careful evaluation for agricultural applications.

4. SIGS technology

The dsRNA can be absorbed through the insect cuticle or digestive system to silence target genes. However, under field conditions, it is challenging for sprayed dsRNA to be directly contacted or ingested by insects. Studies have demonstrated that foliar-applied dsRNA can be distributed throughout the plant via its internal transport system, significantly enhancing the efficacy of pest control [30]. The dsRNA sprayed on leaf surfaces can be detected in roots and newly emerged leaves, with fluorescent signals observable in various cells and structures such as xylem, phloem parenchyma, companion cells, mesophyll cells, trichomes, and stomata. The ability of dsRNA to be absorbed by plants and translocated between leaves allows effective targeting of pests hidden on the undersides of leaves, inside fruits, or within root tissues – areas typically inaccessible to conventional pesticides – thereby increasing the practical relevance of this technology for pest management [31]. By applying nucleic acid pesticides directly to plant surfaces, key genes related to insect development or pathogen infection can be interfered with, reducing the fitness or pathogenicity of pests and ultimately mitigating crop damage.

In recent years, the use of SIGS technology for applying sprayable nucleic acid pesticides to control agricultural pests has gained widespread attention in the field [32]. In 2019, Bayer submitted the world's first sprayable nucleic acid pesticide, BioDirect, to the EPA for approval. This product significantly reduces the survival rate of Varroa destructor, a major pest in beekeeping, without harming honeybees [33]. Spraying dsRNA can effectively induce silencing of target pest genes, circumventing technical limitations and public concerns associated with transgenic plants, thereby offering a novel approach to insect control [34]. Compared to nucleic acid pesticides produced by transgenic plants, sprayable nucleic acid pesticides are favored for their shorter development time, lower cost, reduced resistance risk, straightforward regulation, and applicability to all crops [35]. The ease of application and versatility in various scenarios make sprayable nucleic acid pesticides a primary future application method. Although extensive research has been conducted on RNAi via SIGS against various pests and pathogens, the mechanisms of dsRNA uptake and subsequent cellular internalization in plants are not fully understood [36]. Elucidating these mechanisms is crucial for the large-scale development and optimization of RNAi-based crop protection strategies. Moreover, siRNA and dsRNA molecules exhibit short environmental stability and vary in uptake efficiency among pathogens and pests. Recent advances in nanoparticle-based RNA protection have significantly improved the feasibility of RNA-based crop protection strategies, enhancing the effectiveness of SIGS [37].

Beyond prospective technological demonstrations, empirical data from commercialized products and field trials are gradually revealing the real-world performance of RNAi-based pesticides. For instance, the first commercialized RNAi maize, SmartStax® PRO, which expresses DvSnf7 dsRNA, has demonstrated effective control of the Western Corn Rootworm under field conditions. Data indicate that it can significantly increase larval mortality and reduce the root damage rating by approximately 50%, directly translating into stabilized and improved yield. In spray applications, field trials targeting wheat aphids have shown that nanoparticle-encapsulated dsRNA sprays can achieve over 70% suppression of the target pest population within seven days post-application. However, the data also underscore existing challenges; for example, the efficacy of naked dsRNA against certain lepidopteran pests can decline rapidly due to environmental degradation in the field, with initial control rates potentially reaching 60%, but persistence often lasts less than five days. These real-world data are critical, as they not only validate the potential of RNAi technology but also precisely pinpoint the bottlenecks in environmental stability, species specificity, and cost-effectiveness that require urgent resolution, thereby guiding future technological refinements.

5. Application of nanoparticle-mediated RNAi technology in pest control

Nanoparticle-mediated RNAi technology has demonstrated significant efficacy in the control of various insect pests. In 2013, He et al. [38] reported the use of cationic core-shell fluorescent nanoparticles (FNPs) for delivering dsRNA targeting the chitinase gene CHT10 in the Asian corn borer (*Ostrinia furnacalis*). Larvae fed with an artificial diet containing FNP/CHT10-dsRNA complexes showed suppressed expression of the CHT10 gene, indicating that FNPs can serve as an efficient gene delivery vehicle for dsRNA to silence critical target genes, impede larval development, and ultimately achieve pest control. This study represented the first application of fluorescent nanoparticles for dsRNA delivery in pest management. Subsequently, nanoparticle-mediated dsRNA delivery has been successfully applied in a variety of insect pests. For example, the German cockroach (Blattella germanica) is highly sensitive to injected dsRNA, and RNAi can effectively suppress the expression of target genes. Using liposome-encapsulated dsRNA to delay degradation in the midgut, experiments showed that feeding such dsRNA significantly inhibited α-tubulin gene expression, leading to mortality [39]. Based on previous studies, Yan et al. [40] utilized SPc nanocarriers to deliver dsRNA through topical application and spraying, enabling effective penetration of the cuticle of the soybean aphid (Aphis glycines), resulting in mortality rates of 81.67% and 78.5%, respectively. These successful cases illustrate that the integration of RNAi technology with nanocarriers provides a promising platform for developing new pest control strategies [41]. Recent studies have also shown that SWCNTs can be used for direct delivery of siRNA into plant cells. Owing to their needle-like structure, small diameter (approximately 1 nm), considerable length (around 500 nm), and high rigidity, SWCNTs are capable of traversing the plant cell wall and localizing within cells [42]. They also protect siRNAs from nuclease degradation and efficiently deliver DNA and siRNAs into the cytoplasm, thereby inducing endogenous gene knockdown [43].

6. Challenges to dsRNA stability and corresponding strategies

The dsRNA-based bio-pesticides exhibit great potential in agricultural pest control due to their high target specificity, systemic mobility in crops, and environmentally friendly characteristics. However, a major obstacle to their widespread adoption is the low RNAi efficiency resulting from the inherent instability of dsRNA. The stability of dsRNA is primarily influenced by three categories of factors: the natural environment, the intra-plant environment, and the intra-insect environment.

Despite its considerable potential under laboratory conditions, the practical field application of dsRNA faces a series of formidable physicochemical and biological stability challenges. Firstly, environmental stability is a major hurdle for sprayable dsRNA. Naked dsRNA undergoes rapid photodegradation under natural sunlight, particularly due to ultraviolet (UV) radiation, with a half-life potentially as short as a few hours, severely limiting its field persistence. Secondly, the journey of dsRNA before reaching the target pest is fraught with perils. When applied via foliar spray, dsRNA must withstand variable pH on the plant surface and avoid being washed away by rain or degraded by enzymes from epiphytic microorganisms. Finally, after successful ingestion by the target pest, dsRNA must navigate the formidable barriers of the insect digestive system. The midgut environment (notably the highly alkaline conditions in lepidopteran pests) and the abundant presence of dsRNA-specific nucleases (dsRNases) cause rapid degradation. Furthermore, the peritrophic matrix acts as a physical barrier, hindering the large dsRNA molecules from reaching the intestinal epithelial cells for absorption. Collectively, these factors result in oral delivery, while more direct than topical penetration, exhibiting highly variable and often low efficiency across insect species. Overcoming these interconnected obstacles represents the core challenge for translating dsRNA biopesticides from the lab to the field.

When exposed to natural conditions, naked dsRNA is highly susceptible to degradation, which is a key factor limiting its application via spraying. Environmental factors such as ultraviolet radiation, rainwater wash-off, and microbial activity contribute to its decomposition, and pH also plays a significant role in dsRNA degradation. Although leaf surfaces are generally slightly acidic, interspecific variations in pH exist [44], influenced by leaf physiology and environmental ion availability, which may contribute to differences in RNAi efficiency across crop species. Due to its chemical nature, RNA is generally more stable under acidic than alkaline conditions [45]. Although dsRNA shows some resistance to alkaline hydrolysis, the underlying mechanisms and broader applicability require further investigation.

Nanomaterials serving as delivery vehicles for dsRNA have been proven to enhance RNAi efficiency, primarily by reducing degradation and improving cellular uptake. For instance, nanomaterial-encapsulated dsRNA exhibits improved stability under ambient conditions, resistance to UV degradation, and enhanced adhesion to foliage. Studies demonstrate that dsRNAs in complexes with LDH clay nanosheets (BioClay) are largely retained on leaf surfaces after washing, whereas unprotected dsRNA is readily removed [46]. Another advantage of nanoparticles is their potential to improve the foliar and cellular absorption of sprayed dsRNA [47]. Several strategies have been developed to induce robust RNAi responses by enhancing leaf wettability, cuticle penetration, and solute transport, including physical abrasion of the cuticle, high-pressure treatment, surfactant application, and chemical manipulation of stomatal aperture [48]. It is noteworthy that even when dsRNA is internalized into leaves, it may not necessarily enter plant cells. Insufficient cellular uptake of dsRNA can lead to inadequate siRNA production by the plant RNAi machinery, compromising gene silencing efficacy. Thus, enhancing cellular internalization is critical for effective RNAi [49].

After plant uptake, dsRNA is typically cleaved by Dicer enzymes into 21–24 nt siRNAs, which are inefficiently taken up by insect cells. Although suppressing Dicer activity in transgenic plants can enhance dsRNA stability and insecticidal efficacy, the use of genetically modified crops remains limited for various reasons [50]. Developing non-transgenic methods to improve dsRNA stability in plants is currently a research priority. However, it remains unclear whether nanomaterial-delivered dsRNA can suppress plant Dicer expression in a manner similar to transgenic strategies.

Within insects, dsRNA stability is affected by digestive nucleases, highly alkaline pH, gut microbiota, and the peritrophic matrix [51]. dsRNases present in insect body fluids (e.g., saliva, digestive juice, and hemolymph) rapidly degrade dsRNA and are considered a major factor limiting RNAi efficiency. Reducing dsRNase activity can significantly stabilize dsRNA and improve target gene silencing. For example, in the Colorado potato beetle, midgut fluid exhibits stronger degradative activity than hemolymph; meanwhile, polyethylenimine (PEI)–dsRNA complexes show a protective effect *in vitro*, reducing degradation [52]. Similar phenomena have been observed in other pests, such as the corn rootworm and brown planthopper. Furthermore, ingestion of dsRNA by European honeybees upregulates the expression of SID genes. Silencing LmdsRNase2 in the migratory locust via microinjection reduced dsRNase activity in digestive fluid by 30% and increased target gene interference efficiency by 50%–70% [53]. Gut microbiota can influence pH and secrete nucleases that degrade dsRNA or interfere with cellular uptake. The peritrophic matrix, composed of chitin and glycoproteins, may hinder dsRNA access to intestinal epithelial cells, with pore size affecting dsRNA delivery. Nanomaterials, due to their high affinity and penetration ability, can improve dsRNA delivery across the cuticle or peritrophic membrane, thereby enhancing RNAi efficacy [54].

In summary, the efficiency of dsRNA delivery largely determines the success of RNAi-based pest control. Targeting dsRNase as an auxiliary gene represents a feasible strategy for improving RNAi-mediated mortality. Although multiple delivery methods have been developed – including feeding, transgenic expression, microinjection, soaking, and virus-mediated delivery – none fully meet the demands of field application. Whether spray-induced simultaneous targeting of insect dsRNase and essential genes can enhance pest control efficacy requires further experimental validation [55].

7. The key bridging role of microorganisms in nano-pesticides

In recent years, significant progress has been made in understanding the role of microorganisms, particularly insect endosymbionts, in enhancing RNAi efficiency by modulating the "plant–insect–microbe" interaction network. The important functions of Wolbachia, Serratia marcescens, and other symbiotic bacteria in this regulation have been increasingly revealed [56].

Wolbachia activates the Toll/IMD immune pathway in insects, upregulates antimicrobial peptide expression, and suppresses viral replication (e.g., dengue virus and chikungunya virus) [57]. By competing with viruses for host cell resources (such as cholesterol and nucleotides), it reduces viral load and decreases host nuclease activity, indirectly protecting dsRNA from degradation. For example, in the Aedes albopictus cell line (Aa23), Wolbachia infection reduced viral load by 90% while enhancing dsRNA stability [58]. In the parasitic wasp Trichogramma, Wolbachia induces parthenogenesis and enhances host gene-editing efficiency, as confirmed by RNAi knockdown of its regulatory gene ciB [59]. Transcriptomic analysis of filarial nematodes and their resident Wolbachia via RNase H-based rRNA removal showed that the symbiont suppresses the host RNase III gene (responsible for degrading dsRNA) and upregulates key RNAi genes such as Dicer and Ago2, thereby indirectly enhancing RNAi efficacy [60].

Following the application of ZIF-8@PDA nanocarriers, the abundance of Serratia marcescens in the gut of the fall armyworm (*Spodoptera frugiperda*) increased significantly, while the symbiont *Enterococcus mundtii* was suppressed, leading to microbial dysbiosis and conditioned pathogenicity [61]. The ZIF-8 component of the nanoparticles induced reactive oxygen species (ROS) production in the insect midgut, triggering oxidative stress. *S. marcescens* utilized ROS as a proliferation signal, whereas *E. mundtii*, with weaker antioxidant capacity, was outcompeted. During the immune activation phase (0–48 hours), bacterial invasion activated the Toll/IMD pathway and upregulated Duox, resulting in a burst of ROS aimed at clearing pathogens. In the immunosuppression phase (after 48 hours), S. marcescens secreted prodigiosin to directly inhibit host ROS activity and suppressed immune gene expression by activating the antagonism between juvenile hormone (JH3) and 20-hydroxyecdysone (20E). Immunosuppression compromised the insect's ability to degrade exogenous dsRNA, thereby improving the silencing efficiency of target genes (e.g., chitin synthase CHS and V-ATPase B subunit) delivered by ZIF-8@PDA [62]. An indirect effect was the disruption of midgut tissue structure due to S. marcescens infection, which facilitated nanocarrier penetration and increased the uptake of dsRNA@ZIF-8@PDA by 12.3-fold, synergistically enhancing RNAi efficiency [63].

Other *Serratia* species, such as *Serratia symbiotica*, enhance aphid defense by secreting effector proteins that suppress the predatory behavior of ladybug larvae – for example, reducing their development rate and foraging efficiency, thereby alleviating pressure on aphids [64]. Under symbiont protection, aphids exhibit a reduced stress response and decreased midgut nuclease activity, leading to improved stability of exogenous dsRNA. Field trials showed that aphid populations carrying *S. symbiotica* were 40% more susceptible to dsRNA targeting the ATPase gene, with significantly higher mortality than symbiont-free populations [65]. In Drosophila larvae, digestive enzymes (e.g., amylase and sucrase) activate the symbiotic bacterium Lactobacillus plantarum, which produces lactic acid to inhibit Serratia marcescens [66]. Under lactic acid stress, *S. marcescens* differentiates into high and low-virulence subpopulations, with the low-virulence group reducing host immune response and creating favorable conditions for dsRNA delivery [67].

Lactobacillus plantarum modulates host defense via metabolic regulation. Upon host induction, it increases lactic acid production, lowers the environmental pH, and inhibits pathogen proliferation. Lactic acid also activates the jasmonic acid (JA) signaling pathway in Drosophila intestinal epithelial cells, enhancing antimicrobial peptide expression and reducing gut inflammation, thereby indirectly protecting dsRNA [68]. Single-cell transcriptomic evidence indicates that metabolic heterogeneity within bacterial communities (e.g., glycolysis vs. TCA cycle) influences the host immune status, with a low-inflammatory environment being more conducive to RNAi [69]. In summary, microorganisms such as Wolbachia, Serratia, and Lactobacillus plantarum provide new pathways for overcoming application barriers of RNAi technology in insects by reshaping the host immune–metabolic network. Future research should focus on species specificity, ecological safety, and field application efficiency [70].

8. Conclusion

While RNAi technology presents a revolutionary prospect for green pest management, bridging the gap from laboratory promise to widespread field application necessitates overcoming significant hurdles. Technologically and economically, the current high cost of industrial-scale dsRNA production directly impacts end-product affordability and accessibility, potentially excluding resource-limited smallholder farmers. From regulatory and social perspectives, the absence of a globally harmonized regulatory framework for RNAi-based products, coupled with evolving approval processes, biosafety assessment standards, and long-term environmental risk evaluation systems, creates uncertainty for commercial rollout. More critically, the assertion of "high specificity and environmental safety" demands careful scrutiny. Growing evidence highlights non-negligible ecological risks, including off-target effects, potential impacts on non-target organisms (e.g., beneficial insects and soil microbiota), and the environmental fate of dsRNA. Therefore, future advancement must place comprehensive risk assessment, inclusive socio-economic analysis, and internationally coordinated regulatory development at its core, alongside technological innovation for cost reduction and efficacy improvement. Only through interdisciplinary collaboration can this powerful technology be ensured to serve the global agricultural system in a safe, equitable, and sustainable manner.

Funding

This research was funded by the Bingtuan Science and Technology Program of XPCC (2024DA018), the National Natural Science Foundation of China (32260676) and Supported by the earmarked fund for XARS(2026XARS-04-202610).

Author details

Yiying Zhao*, Xiaojun Wang, Kedi Zhao, Qiancheng Wei and Feng Liu
College of Agriculture, Shihezi University, Shihezi, China

*Address all correspondence to: zhaoyiying@shzu.edu.cn

References

[1] Tay WT, Meagher RL, Groot AT. Spodoptera frugiperda: Ecology, evolution, and management options of an invasive species. Annual Review of Entomology. 2023;**68**:299–317

[2] Mendesil E, Tefera T, Hutchison WD. The invasive fall armyworm, Spodoptera frugiperda, in Africa and Asia: Responding to the food security challenge, with priorities for integrated pest management research. Journal of Plant Diseases and Protection. 2023;**130**:1175–1206

[3] Li Y. Managing the invasive fall armyworm through biotech crops: A Chinese perspective. Trends in Biotechnology. 2021;**39**:105–107

[4] Lu J, Shen J. Target genes for RNAi in pest control: A comprehensive overview. Entomologia Generalis. 2024;**44**:95–114

[5] Zhu KY, Palli SR. Mechanisms, applications, and challenges of insect RNA interference. Annual Review of Entomology. 2020;**65**:293–311

[6] Meister G, Tuschl T. Mechanisms of gene silencing by double-stranded RNA. Nature. 2004;**431**:343–349

[7] Hutvagner G, Simard MJ. Argonaute proteins: Key players in RNA silencing. Nature Reviews Molecular Cell Biology. 2008;**9**:22–32

[8] Tang W, Luo XY, Sanmuels V. Gene silencing: Double-stranded RNA mediated mRNA degradation and gene inactivation. Cell Research. 2001;**11**:181–186

[9] Dalakouras A, Koidou V, Papadopoulou K. DsRNA-based pesticides: Considerations for efficiency and risk assessment. Chemosphere. 2024;**352**:141530

[10] Wang Y, Li M, Wang X. High-efficiency green management of potato late blight by a self-assembled multicomponent nano-bioprotectant. Nature Communications. 2023;**14**:1–10

[11] Yoon J-S, Gurusamy D, Palli SR. Accumulation of dsRNA in endosomes contributes to inefficient RNA interference in the fall armyworm, Spodoptera frugiperda. Insect Biochemistry and Molecular Biology. 2017;**90**:53–60

[12] Dahlman JE, Barnes C, Anderson DG. In vivo endothelial siRNA delivery using polymeric nanoparticles with low molecular weight. Nature Nanotechnology. 2014;**9**:648–655

[13] Yu C, Li J, Zhang Z, He S. Metal–organic framework-based insecticide and dsRNA codelivery system for insecticide resistance management. ACS Applied Materials & Interfaces. 2023;**15**:48495–48505

[14] Jain RG, Fletcher SJ, Mitter N. Foliar application of clay-delivered RNA interference for whitefly control. Nature Plants. 2022;**8**:535–548

[15] Zhou H, Wan F, Jian Y, Ding W. Chitosan/dsRNA polyplex nanoparticles advance environmental RNA interference efficiency through activating clathrin-dependent endocytosis. International Journal of

Biological Macromolecules. 2023;**253**:127021

[16] Zhang X, Zhang J, Zhu KY. Chitosan/double-stranded RNA nanoparticle-mediated RNA interference to silence chitin synthase genes through larval feeding in the African malaria mosquito (Anopheles gambiae). Insect Molecular Biology. 2010;**19**:683–693

[17] Theresa M, Allen TM, Cullis PR. Liposomal drug delivery systems: From concept to clinical applications. Advanced Drug Delivery Reviews. 2013;**65**:36–48

[18] Castellanos NL, Smagghe G, Sharma R, Oliveira EE, Christiaens O. Liposome encapsulation and EDTA formulation of dsRNA targeting essential genes increase oral RNAi-caused mortality in the Neotropical stink bug Euschistus heros. Pest Management Science. 2019;**75**:537–548

[19] Yan S, Liu X, Shen J. Nanoparticle-mediated double-stranded RNA delivery system: A promising approach for sustainable pest management. Insect Science. 2021;**28**:21–34

[20] Jiang Z, Zhang J, Bock R. Next-generation insect-resistant plants: RNAi-mediated crop protection. Trends in Biotechnology. 2017;**35**:871–882

[21] Ocsoy I, et al. Nanotechnology in plant disease management: DNA-directed silver nanoparticles on graphene oxide as an antibacterial against Xanthomonas perforans. ACS Nano. 2013;**7**:8972–8980

[22] Dhandapani RK, Gurusamy D, Palli SR. Protamine–Lipid–dsRNA nanoparticles improve RNAi efficiency in the fall armyworm, spodoptera frugiperda. Journal of Agricultural and Food Chemistry. 2022;**70**:6634–6643

[23] Chao Z, Ma Z, Shen J. Establishment of star polycation-based RNA interference system in all developmental stages of fall armyworm Spodoptera frugiperda. Entomologia Generalis. 2023;**43**:127–137

[24] Mason CJ, Peiffer M, Felton GW. Concerted impacts of antiherbivore defenses and opportunistic Serratia pathogens on the fall armyworm (Spodoptera frugiperda). Oecologia. 2022;**198**:167–178

[25] Gomes AFF, Omoto C, Cônsoli FL. Gut bacteria of field-collected larvae of Spodoptera frugiperda undergo selection and are more diverse and active in metabolizing multiple insecticides than laboratory-selected resistant strains. Journal of Pest Science. 2020;**93**:833–851

[26] Chung SH, Rosa C, Felton GW. Herbivore exploits orally secreted bacteria to suppress plant defenses. Proceedings of the National Academy of Sciences. 2013;**110**:15728–15733

[27] Sahu KP, Singh S, Singh SK. Nanomaterials via ZIF-8: Preparations, catalytic and drug delivery applications. Chemical Engineering Journal (Lausanne, Switzerland : 1996). 2025;**508**:160663

[28] Li H, Chen W, Liu B, Chen G. A purely green approach to low-cost mass production of zeolitic imidazolate frameworks. Green Energy & Environment. 2023;**8**:775–784

[29] Tong Y, Shao L, Wu X. Adhesive and stimulus-responsive polydopamine-coated graphene oxide system for pesticide-loss control.

Journal of Agricultural and Food Chemistry. 2018;**66**:2616–2622

[30] Gao Z, Tan J, Sun Y, Jiang X. Size effect of ZIF -8 based nanocarrier pesticide delivery system on targeted release and insecticidal activity. Pest Management Science. 2025;**81**:966–977

[31] Altschul SF, et al. Basic local alignment search tool. Journal of Molecular Biology. 1990;**215**:403–410

[32] Andrews S FastQC. GitHub Repository. 2023

[33] Anvari D, et al. The global status of Dirofilaria immitis in dogs: A systematic review and meta-analysis. Research in Veterinary Science. 2020;**131**:104–116

[34] Bandi C, et al. Phylogeny of Wolbachia in filarial nematodes. Proceedings of the Royal Society of London. Series B: Biological Sciences. 1998;**265**:2407–2413

[35] Bazzocchi C, et al. Combined ivermectin and doxycycline treatment has microfilaricidal and adulticidal activity against Dirofilaria immitis in experimentally infected dogs. International Journal for Parasitology. 2008;**38**:1401–1410

[36] Bhalla D, Dumas M, Preux PM. Neurological manifestations of filarial infections. In: Garcia HH, Tanowitz HB, Del Brutto OH, editors. Handbook of Clinical Neurology. Elsevier; 2013

[37] Boussinesq ML. Loiasis. Annals of Tropical Medicine & Parasitology. 2006;**100**:715–731

[38] He S, et al. Validation of two ribosomal RNA removal methods for microbial metatranscriptomics. Nature Methods. 2010;7:807–812

[39] Hoerauf A, et al. Wolbachia endobacteria depletion by doxycycline as antifilarial therapy has macrofilaricidal activity in onchocerciasis: A randomized placebo-controlled study. Medical Microbiology and Immunology. 2008;**197**:295–311

[40] Yan S, et al. Spray method application of transdermal dsRNA delivery system for efficient gene silencing and pest control on soybean aphid Aphis glycines. Journal of Pest Science. 2020;**93**:449–459

[41] Kang Y, et al. Transcript amplification from single bacterium for transcriptome analysis. Genome Research. 2011;**21**:925–935

[42] Kim D, et al. Graph-based genome alignment and genotyping with HISAT2 and HISAT-genotype. Nature Biotechnology. 2019;**37**:907–915

[43] Kumar N, et al. Efficient subtraction of insect rRNA prior to transcriptome analysis of Wolbachia-Drosophila lateral gene transfer. BMC Research Notes. 2012;**5**:230

[44] Li H, et al. The sequence alignment/Map format and SAMtools. Bioinformatics. 2009;**25**:2078–2079

[45] Li Z, Carlow CKS. Characterization of transcription factors that regulate the type IV secretion system and riboflavin biosynthesis in Wolbachia of Brugia malayi. PLoS ONE. 2012;7:e51597

[46] Liao Y, Smyth GK, Shi W. featureCounts: An efficient general purpose program for assigning sequence reads to genomic features. Bioinformatics. 2014;**30**:923–930

[47] Luck AN, et al. Heme acquisition in the parasitic filarial nematode Brugia malayi. The FASEB Journal. 2016;**30**:3501–3514

[48] Mackenzie CD, et al. Approaches to the control and elimination of the clinically important filarial diseases. In: Klei TR, Rajan TV, editors. The Filaria. Springer; 2002. p. 155–165

[49] Mand S, et al. Macrofilaricidal activity and amelioration of lymphatic pathology in bancroftian filariasis after 3 weeks of doxycycline followed by single-dose diethylcarbamazine. The American Journal of Tropical Medicine and Hygiene. 2009;**81**:702–711

[50] Martin M. Cutadapt removes adapter sequences from high-throughput sequencing reads. EMBnet.journal. 2011;**17**:10–12

[51] Mclaren DJ, et al. Micro-organisms in filarial larvae (Nematoda). Transactions of the Royal Society of Tropical Medicine and Hygiene. 1975;**69**:509–514

[52] Medeiros ZM, et al. Lymphatic filariasis: A systematic review on morbidity and its repercussions in countries in the Americas. International Journal of Environmental Research and Public Health. 2021;**19**:316

[53] Mediannikov O, Ranque SM. the most neglected human filariasis. New Microbes and New Infections. 2018;**26**: S19–S22

[54] Michalski ML, et al. The NIH-NIAID filariasis research reagent resource center. PLoS Neglected Tropical Diseases. 2011;**5**:e1261

[55] Molyneux DH, et al. Mass drug treatment for lymphatic filariasis and onchocerciasis. Trends in Parasitology. 2003;**19**:516–522

[56] Parkinson J, et al. A transcriptomic analysis of the phylum Nematoda. Nature Genetics. 2004;**36**:1259–1267

[57] Prezza G, et al. Improved bacterial RNA-seq by Cas9-based depletion of ribosomal RNA reads. RNA. 2020;**26**:1069–1078

[58] Quek S, et al. Wolbachia depletion blocks transmission of lymphatic filariasis by preventing chitinase-dependent parasite exsheathment. Proceedings of the National Academy of Sciences. 2022;**119**:e2120003119

[59] Lu YH, Wu KM. Mirid bugs in China: Pest status and management strategies. Outlooks on Pest Management. 2011;**22**:248–252

[60] Lu YH, et al. Mirid bug outbreaks in multiple crops correlated with wide-scale adoption of Bt cotton in China. Science. 2010;**328**:1151–1154

[61] Tan YA, et al. Phospholipase C gamma (PLCγ) regulates soluble trehalase in the 20E-induced fecundity of Apolygus lucorum. Insect Science. 2021;**28**:430–444

[62] Zamore PD, et al. RNAi: Double-stranded RNA directs the ATP-dependent cleavage of mRNA at 21 to 23 nucleotide intervals. Cell. 2000;**101**:25–33

[63] Minakuchi C, Zhou X, Riddiford LM. Krüppel homolog 1 (Kr-h1) mediates juvenile hormone action during metamorphosis of Drosophila melanogaster. Mechanisms of Development. 2008;**125**:91–105

[64] Daimon T, et al. Knockout silkworms reveal a dispensable role for juvenile hormones in holometabolous life cycle. Proceedings of the National Academy of Sciences. 2015;**112**: E4226–E4235

[65] Heigwer F, Port F, Boutros M. RNA interference (RNAi) screening in Drosophila. Genetics. 2018;**208**:853–874

[66] Zhu F, et al. A brain-specific cytochrome P450 responsible for the majority of deltamethrin resistance in the QTC279 strain of Tribolium castaneum. Proceedings of the National Academy of Sciences. 2010;**107**: 8557–8562

[67] Zhu F, et al. Bed bugs evolved unique adaptive strategy to resist pyrethroid insecticides. Scientific Reports. 2013;**3**:1456

[68] Bai H, Palli SR. Identification of G protein-coupled receptors required for vitellogenin uptake into the oocytes of the red flour beetle, Tribolium castaneum. Scientific Reports. 2016;**6**:27648

[69] Baum JA, et al. Control of coleopteran insect pests through RNA interference. Nature Biotechnology. 2007;**25**:1322–1326

[70] Poreddy S, Li J, Baldwin IT. Plant-mediated RNAi silences midgut-expressed genes in congeneric lepidopteran insects in nature. BMC Plant Biology. 2017;**17**:1–10

Chapter 7

The Melon Fly (*Myiopardalis Pardalina*) (Diptera: Tephritidae): Identification, Biology, Damage Characteristics, and Management Strategies

Aydemir Bariş

Abstract

Myiopardalis pardalina (Bigot, 1891) (Diptera: Tephritidae), commonly known as the Baluchistan melon fly, is an invasive pest of growing economic significance, particularly in cucurbit-producing regions of Asia, the Middle East, and Türkiye. This review synthesizes current knowledge on the species' taxonomy, biology, host range, damage mechanisms, geographic distribution, and control strategies. *M. pardalina* undergoes complete metamorphosis and is most destructive during its larval stage, when it tunnels through the fruit mesocarp, causing tissue collapse, secondary microbial infections, and reduced market value. Although its host range is narrow – mainly limited to *Cucumis melo* and *Citrullus lanatus* – its adaptability to both open-field and greenhouse environments exacerbates its pest status. Recent studies indicate that climate change, monoculture practices, and expanded trade have facilitated the pest's spread and accelerated its generational turnover. Cultural control measures – including sanitation, synchronized planting, and physical barriers – have proven effective in suppressing populations. Natural enemies such as predatory ants (*Cataglyphis* spp., *Pheidole* spp.), entomopathogenic fungi (*Beauveria bassiana*, *Metarhizium anisopliae*), and entomopathogenic nematodes show promise for integration into environmentally sustainable pest-management programs; however, specific parasitoids targeting *M. pardalina* have yet to be identified. Chemical control is considered a secondary strategy, with precise timing essential for optimal results. By consolidating ecological, biological, and management-related data on *M. pardalina* and highlighting critical knowledge gaps, this review aims to support the development of integrated pest-management (IPM) programs.

Keywords: melon fly, *Myiopardalis pardalina*, biology, damage mechanism, integrated pest management, natural enemies, cultural control measures.

1. Introduction

The melon fly, or Baluchistan melon fly (*Myiopardalis pardalina*) (*M. pardalina*), a member of the family Tephritidae (order Diptera), has become a highly destructive and invasive pest that poses a serious threat to cucurbit crops worldwide. Especially prevalent in parts of Asia, Africa, and Europe, *M. pardalina* is linked to catastrophic economic losses, with crop damage occasionally exceeding 90% during severe outbreaks. Its strong preference for species in the Cucurbitaceae family – including melon (*Cucumis melo*), watermelon (*Citrullus lanatus*), and cucumber (*Cucumis sativus*) – makes it a major constraint on the sustainable cultivation of these crops. The rising frequency and intensity of infestations in Türkiye, India, Iran, Pakistan, and the wider Middle East highlight the urgent need for effective, science-based control strategies.

A detailed understanding of the pest's biology and ecology is essential for reducing its impact. *M. pardalina* undergoes complete metamorphosis – egg, larva, pupa, and adult – with each stage displaying distinct morphological and physiological traits that confer survival advantages across diverse environments. Larvae inflict the greatest damage, feeding on the fruit mesocarp and creating galleries that cause tissue breakdown, microbial contamination, and marked declines in yield and market value. The pest's ability to thrive in both open-field and protected cultivation systems, coupled with growing tolerance to fluctuating climates, further complicates management.

Recent research underscores the value of studying the fly's behavioral ecology – mating, oviposition, and dispersal – in order to develop more targeted control measures. International trade has been identified as a major pathway for its spread, underscoring the need for robust quarantine and phytosanitary protocols. Chemical attractants and bait systems, together with biological agents such as parasitoids and entomopathogenic fungi or nematodes, are being explored as alternative or complementary components of integrated pest-management (IPM) programs. Cultural practices – including field sanitation, crop rotation, synchronized planting, and physical exclusion – have likewise shown promise in suppressing populations and reducing fruit infestation.

This chapter offers a comprehensive overview of the biology, damage mechanisms, natural enemies, geographic distribution, and management of *M. pardalina*. Special emphasis is placed on the pest's potential for further range expansion under global climate change and the associated risks to agricultural systems. Integrating ecological insights with practical management tools is vital for developing sustainable, region-specific strategies that can contain this economically important pest and safeguard cucurbit production in affected areas.

2. Definition and biology

The melon fly *M. pardalina* (Bigot, 1891) (Diptera: Tephritidae) is a major pest of melons and other cucurbits, causing substantial yield and quality losses in high-value field crops. In Türkiye, the species was first reported by Alkan (1946) and Bodenheimer (1958) [1, 2]. Rising temperatures in recent years have broadened its distribution, increasing the threat to growers. *M. pardalina* has been documented in Afghanistan, Cyprus, India, Iraq, Israel, Lebanon,

Pakistan, Persia (Iran), Syria, Türkiye, the former USSR, several African countries, and Senegal [3, 4].

A sound understanding of the fly's biology is essential for effective management. Adults become active in summer and typically oviposit in young melon fruits roughly the size of a hazelnut. The hatched larvae feed within the mesocarp, creating tunnels (larval galleries) as they progress toward the locular cavity (seed chamber). After completing their feeding phase, mature larvae drop to the soil to pupate. Under Ankara's ecological conditions, the species completes two to three generations per year, with development from egg to adult lasting 36–39 days [5].

Morphologically, *M. pardalina* displays the typical traits of tephritid flies [6]. Females are larger than males and bear distinctive yellow wing bands and three-segmented antennae. Eggs are translucent white and measure about 1.2–1.3 mm in length. Both morphological and behavioral differences occur among larval instars. During pupation, the larva forms a barrel-shaped puparium in the soil. Detailed characterization of these features is vital for designing IPM strategies (**Figures 1–3**).

The average female wingspan is 5.08 ± 0.04 mm, and body length is 6.67 ± 0.06 mm; in males, these dimensions are 4.18 ± 0.03 mm and 5.68 ± 0.06 mm, respectively [6]. Pupae are usually found within the top 1–6 cm of soil, although

Figure 1.
(A) Female melon fly, (B) male.

Figure 2.
(A) Mature larva of the melon fly, (B) mature larva leaving the melon to pupate in the soil.

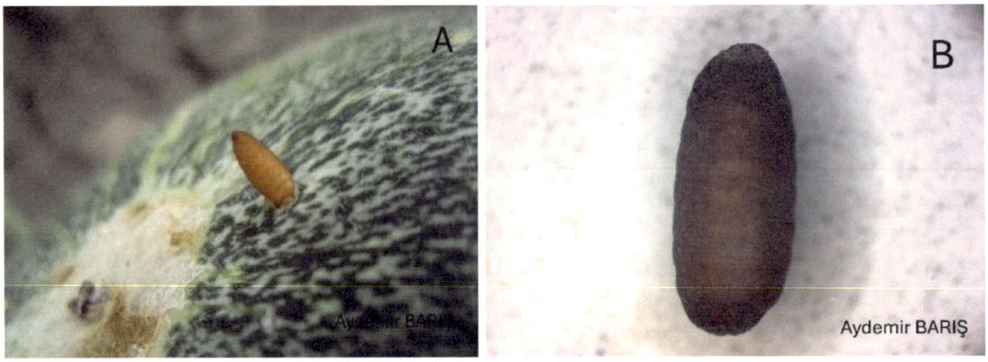

Figure 3.
(A, B), Pupa of the melon fly.

depths of 15–16 cm have been reported. Neonate larvae typically begin feeding near the fruit surface, later moving toward the seed cavity and, once the fruit ripens, exiting to pupate in the soil.

3. Damage and economic importance

Damage caused by the melon fly, *M. pardalina*, is concentrated in the larval stage. Females deposit their eggs just beneath the epidermis of young fruits roughly the size of a hazelnut. To oviposit, the female punctures the fruit with her ovipositor, leaving tiny holes (**Figure 4**). Sap exuding from these punctures can stain the skin reddish, although the marks often fade as the fruit matures, and adults occasionally feed on the ooze. Mating begins soon after adult emergence: males grasp females firmly and may continue copulation while flying short distances between plants (**Figure 5**).

A single female may lay multiple eggs in the same fruit, and larvae of different instars often coexist, suggesting repeated oviposition either by one female over time or by several females on a shared host. Newly hatched larvae feed in the mesocarp, excavating linear or irregular galleries. Early instars remain near the surface, whereas

Figure 4.
Holes where the melon fly lays eggs in a hazelnut-sized fruit, (B) discharge from the fruit tissue when the female inserts her ovipositor.

Figure 5.
(A) Mating individuals of the melon fly, (B) short-distance flight of individuals together.

later instars migrate toward the seed cavity, ultimately causing internal collapse, discoloration, and foul odors (**Figure 6**) [5].

Larval development can be completed in fruits of various sizes. Eggs are deposited in hazelnut-sized fruits (**Figure 7A**), and as the fruit enlarges, the larvae progress through successive instars within it. On melon plants, fruits ranging from hazelnut to walnut size – and even larger – can support full larval development (**Figure 7B**). Females continue to oviposit in walnut-sized or larger fruits until the pericarp becomes too thick to penetrate; if it later softens due to physiological or environmental reasons, oviposition can resume. In hazelnut-sized fruits, larvae may consume nearly all the mesocarp, leaving only the peel, and several individuals can develop simultaneously (**Figure 8**).

Larval feeding results in the decay of internal fruit tissues, leading to the development of brown lesions and the accumulation of frass. The feeding tunnels cause physical deformation and create entry points for secondary pathogens, particularly saprophytic fungi. This accelerates fruit rot, produces foul odors, and impairs flavor. These symptoms become especially pronounced as the fruit ripens. In severely infested fruits, the entire seed cavity and surrounding tissues may be destroyed. Browning of all tissues within the seed cavity, along with the presence of

Figure 6.
(A) Young larvae feeding in the fleshy tissue of the fruit, (B) damage caused by larval feeding in the fruit's pith.

Figure 7.
(A) Laying eggs on a hazelnut-sized melon fruit, (B) laying eggs on a walnut-sized or larger fruit.

Figure 8.
(A) Damage caused by larvae feeding on hazelnut-sized fruit, (B) blackening of the fruit flesh, leaving only the peel due to larval feeding.

larvae and their droppings in the galleries, significantly reduces both the quality and market value of the fruit for producers [7].

After completing their development, larvae usually leave the fruit in the evening or at night by boring exit holes and dropping to the soil to pupate. The size and shape of these holes vary with the fruit's stage of ripeness. When several larvae inhabit the same fruit, some may reuse and enlarge existing tunnels, so a single fruit can display exit holes of multiple diameters (**Figure 9**).

Emergence is difficult in fruits with a hardened rind: some larvae become trapped while exiting or fail to pupate, dying either inside the fruit or at the exit site. Occasionally, fully developed larvae form a puparium inside the fruit itself, preventing adult emergence and breaking the life cycle. The exit wounds also attract saprophagous flies, which lay eggs in the damaged tissue and hasten decay. Once the rind is perforated, rot progresses rapidly, and the feeding activity of these secondary fly larvae can render even a lightly infested fruit unmarketable (**Figure 10**).

Damage caused by *M. pardalina* is not always externally visible, particularly during the early larval stages. However, applying gentle pressure to the fruit may

Figure 9.
(A, B) Exit holes made by mature melon fly larvae on the fruit to pupate.

Figure 10.
(A) Healthy and damaged fruit, (B) complete rotting of the fruit due to melon fly damage.

reveal internal softening or hollow areas resulting from tissue breakdown. In some cases, only localized yellowing or deformities are observed. These symptoms tend to be more pronounced in specific melon cultivars. For instance, in the "İpsala" variety, localized yellowing associated with larval feeding is a characteristic symptom (**Figure 11**). In other cultivars, external signs of infestation may be absent even when larvae are present. At low infestation levels, the damage remains localized; however, at high larval densities, the entire seed cavity can be destroyed [8].

Field observations indicate that some melon plants bear fruits at multiple developmental stages simultaneously, providing continuous opportunities for oviposition throughout the growing season. Certain fruits stop growing once they reach walnut size or slightly larger, yet still serve as favorable sites for oviposition and larval development. Deformities or softened areas on such fruits further increase their attractiveness for egg-laying. As a result, damage caused by *M. pardalina* not only leads to yield losses but also complicates harvest scheduling and the timing of chemical control applications.

Figure 11.
(A) Exit hole of the pest and the surrounding yellowing in the Ipsala variety, (B) the type of damage occurring in fruits of different sizes.

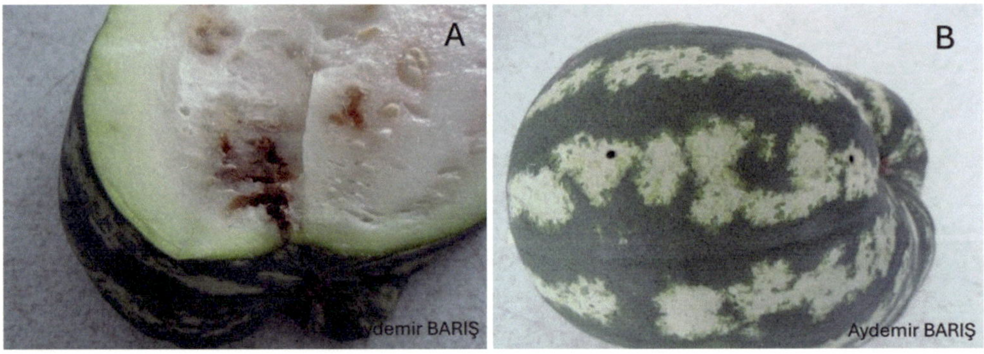

Figure 12.
(A, B) Symptoms of melon fly damage on watermelon.

The nature of the damage differs between melons and watermelons. While *M. pardalina* larvae typically tunnel toward the seed cavity in melons, in watermelons, the damage is usually localized around the feeding site. This results in tissue necrosis and deformation at those specific points (**Figure 12**).

4. Geographic distribution

The melon fly, *M. pardalina*, is originally a Central Asian species that has progressively expanded its geographic range. In Türkiye, it was first recorded by Bodenheimer (1941) and Alkan (1946) and has since been reported to cause economic damage and population outbreaks in provinces such as Elazığ, Niğde, and Konya. At present, the pest is commonly found in the Southeastern Anatolia, Central Anatolia, and Aegean regions. Nevertheless, its overall distribution within Türkiye is still considered relatively limited.

According to the European and Mediterranean Plant Protection Organization (EPPO), *M. pardalina* has been documented in Afghanistan, India, Iran, Iraq, Israel,

Lebanon, Syria, Cyprus, and various former Soviet republics – particularly those in Central Asia. Its presence has also been reported in Georgia, Ukraine, and China. The species prefers warm, semi-arid climates and has adapted well to melon-growing regions, allowing it to establish viable populations across a wide range of agro-ecological zones.

In recent years, rising temperatures associated with climate change have altered the pest's life cycle, often resulting in increased generational turnover and expanded distribution. Intensive monoculture melon cultivation has further facilitated its spread by ensuring a continuous supply of host material. In addition, the transport of infested fruits between regions represents a significant pathway for dispersal. Understanding the distribution patterns of *M. pardalina* is essential for developing effective, region-specific IPM strategies.

5. Host range

The melon fly (*M. pardalina*) is a highly host-selective pest, primarily associated with plants in the family Cucurbitaceae. Within this family, its most important and preferred hosts are melon (*Cucumis melo* L.) and snake melon (*Cucumis melo* var. *flexuosus*). The fruit morphology and cultivation practices of these species provide highly suitable conditions for female oviposition and larval development.

Field observations have shown that, in addition to these primary hosts, *M. pardalina* can also develop on watermelon (*Citrullus lanatus*), although the nature of the damage differs. On watermelon, oviposition and larval activity are generally limited to the rind, resulting in external deformities and irregular hardening of the mesocarp. These symptoms are more pronounced in immature fruits and typically manifest as corky or fibrous lesions [6].

The pest is capable of egg-laying not only in immature fruits at the hazelnut stage but also in fruits that have reached walnut size or slightly larger, even if fruit development has ceased. In such cases, larvae can still complete their development successfully. Some fruits may appear externally healthy while harboring developing larvae inside, indicating that host suitability depends not only on plant species but also on the physiological condition of the fruit.

According to EPPO and FAO sources, the primary host plants of *M. pardalina* include:

Cucumis melo (melon)

Cucumis melo var. *flexuosus* (snake melon)

Citrullus lanatus (watermelon)

Occasionally, *Lagenaria siceraria* (bottle gourd) and other cucurbits [2, 6].

Although the host range is relatively narrow, it is highly effective, which has important implications for biological control and IPM planning. Furthermore, local cultivars – such as "İpsala" and "Kırkağaç" – have been observed to exhibit varying levels of susceptibility, making varietal resistance an important consideration for future breeding programs and cultivar selection.

6. Natural enemies

Control of *M. pardalina* in natural ecosystems is limited due to the scarcity of known natural enemies. However, field observations have identified certain ant species as effective predators. In Türkiye, surveys have shown that *Cataglyphis bicolor* (Fabricius), *Cataglyphis magalocola* (Förster), and *Pheidole pallidula* (Nylander) (Hymenoptera: Formicidae) prey on *M. pardalina* larvae or pupae as they emerge onto the soil surface. These ants forage beneath fruits and can significantly reduce pupation success, thereby exerting suppressive pressure on pest populations [6].

To date, no parasitoid species has been conclusively identified as a specific natural enemy of *M. pardalina*. However, research on other Tephritidae suggests potential for parasitoids as biological control agents. For example, *Pteromalus coloradensis* (Hymenoptera: Pteromalidae) and *Paracantha gentilis* are known to parasitize larvae and pupae of various tephritid species [9]. *P. coloradensis* is especially common in North America, occurring in the floral heads of *Cirsium* (thistles), and is considered a promising candidate for augmentative biological control.

Another notable parasitoid, *Pteromalus* nr. *myopitae* (Hymenoptera: Pteromalidae), has demonstrated effectiveness against the pupae of economically important tephritids, such as the olive fruit fly (*Bactrocera oleae*). This species is widespread throughout the Mediterranean and may possess the potential to parasitize soil-pupating tephritids like *M. pardalina*.

Among entomopathogenic fungi, *Beauveria bassiana* (Balsamo) (Cordycipitaceae) has shown high larvicidal efficacy against tropical tephritids, such as *Anastrepha fraterculus* [10]. Laboratory bioassays report mortality rates of up to 80% in larvae. Similarly, *Metarhizium anisopliae* (Metschnikoff) (Clavicipitaceae) is a well-known soilborne entomopathogen commonly tested for its effectiveness against Tephritidae pests [11].

The conservation and augmentation of these natural enemies in agricultural landscapes are essential for the success of IPM programs. In particular, these biological agents offer sustainable alternatives or supplements to chemical control, especially in organic or environmentally sensitive farming systems. Therefore, systematic monitoring and potential integration of *M. pardalina*'s natural enemies into biological control programs will contribute to the long-term suppression of this pest species.

6.1. Cultural control measures

Cultural control methods aim to disrupt the pest's life cycle by targeting its reliance on environmental conditions, thereby reducing its population. Based on original field observations and practical applications, the following measures have proven effective:

6.1.1. Removal and burial of infested fruits

During field inspections, fruits showing signs of larval infestation – such as entry holes, feeding galleries, or frass – should be collected and buried in deep pits. These pits are then treated with lime and covered with soil. This practice prevents larvae from reaching the soil surface to pupate, thereby interrupting the development of the next generation.

6.1.2. Early intervention

Fruits exhibiting one or two oviposition punctures should be removed promptly. These entry points indicate that larvae are maturing and likely to exit the fruit soon to pupate in the soil. If such fruits remain in the field, they can serve as sources of infestation.

6.1.3. Regionally adapted cultivation practices

In regions like Van (Türkiye), covering melon fruits with soil has proven effective in preventing female flies from accessing the fruit for oviposition. This physical barrier can eliminate the need for second or third pesticide applications. Another protective technique involves covering fruits at the hazelnut stage with mesh or tulle, which allows light and air penetration while preventing oviposition. The covering can be maintained until the fruit reaches a size less susceptible to egg-laying. In greenhouse cultivation, closing greenhouse entrances with mesh barriers has also been shown to prevent *M. pardalina* infestation.

6.1.4. Synchronized production planning

The presence of fruits at different developmental stages within the same field provides *M. pardalina* with continuous opportunities for oviposition. Because melon plants produce flowers and fruits simultaneously throughout the growing season, the pest benefits from prolonged host availability. Collecting immature fruits ranging from hazelnut to walnut size is important for population suppression. Therefore, cultivation techniques that promote synchronous fruit development and harvesting are recommended. Additionally, the use of early-maturing or late-maturing cultivars can help reduce the pest's impact.

6.1.5. Field sanitation and post-harvest clean-up

After harvest, all plant residues – including damaged fruits, rinds, and fallen fruit – should be removed from the field, as they can harbor larvae. Furthermore, deep tillage after harvest can increase the mortality of soil-dwelling pupae.

6.1.6. Post-harvest fruit inspection

Harvested fruits should be individually examined for signs of infestation. Infected fruits must be discarded to prevent the spread of the pest from infested to pest-free areas.

6.2. Chemical control

For effective and environmentally conscious chemical control of *M. pardalina*, precise timing of pesticide applications and the use of registered, appropriate formulations are essential.

6.2.1. Timing of applications

The first pesticide application should be carried out when melon fruits reach the early developmental stage – approximately 0.7 cm in diameter (hazelnut-sized). This stage is critical because it coincides with the onset of oviposition by adult females. Missing this timing may compromise control efforts. Therefore, fields should be monitored regularly starting from the flowering stage, and the first insecticide treatment should be initiated as soon as early fruit set is observed.

A second application should be performed approximately 15 days after the initial treatment. In cases where the cultivation period is extended or fruit development is asynchronous, a third application may be necessary. This is because the pest continues to oviposit in fruits ranging from the hazelnut to walnut stage, and even in larger, fully grown but physiologically arrested fruits. Larvae are capable of completing their development in such fruits as well. Recognizing that oviposition is not restricted to hazelnut-sized fruits is critical for determining optimal treatment windows.

7. Conclusion and recommendations

The melon fly (*M. pardalina*) is a major pest of cucurbit crops, particularly melons, in Türkiye and other warm-climate regions. It causes significant yield and quality losses by tunneling within fruit tissues, leading to secondary infections and reduced market value. The damage is especially severe in melons. Surveys have shown that all melon cultivars grown in Türkiye are susceptible. Even in fields where both melon and watermelon are cultivated, *M. pardalina* demonstrates a strong preference for melons – even under high population pressure [5]. Understanding the pest's biology, behavioral traits, and host preferences is essential for the effective implementation of IPM strategies.

This study synthesizes both literature findings and long-term field observations to provide a holistic view of *M. pardalina*'s life cycle, damage mechanisms, distribution, natural enemies, and control approaches. Biological characteristics, such as oviposition in hazelnut-sized fruits, larval development in fruits at various stages, and pupation in the soil, complicate pest control efforts. Additionally, the pest's population density and distribution range are expanding under the influence of climate change and modern agricultural practices. In areas experiencing milder winters due to global warming, lower winter mortality rates during the pupal stage have been observed. Conversely, increased mortality may occur in regions experiencing harsher winter conditions.

To enhance the effectiveness of *M. pardalina* management, the following recommendations are proposed:

Implement area-wide monitoring systems to track seasonal population dynamics, and base control decisions on these data.

Effectively manage the pest through the full implementation of cultural control practices, systematic field monitoring throughout the growing season, prompt removal of infested melons at the initial stage of infestation (1–2 oviposition holes), and improved grower awareness regarding sanitation and field hygiene.

Promote biological control research, especially on predatory ants, potential parasitoids, and entomopathogenic fungi and nematodes, through systematic evaluations.

Evaluate and utilize local cultivars with demonstrated tolerance to infestation; promote the cultivation of resistant or less-susceptible varieties.

Restrict chemical control to essential cases only, ensuring the use of approved plant protection products during recommended periods.

In conclusion, successful management of the melon fly requires a multifaceted approach, integrating sustainable, environmentally friendly control measures. Enhancing grower awareness and supporting research and policy initiatives will contribute significantly to mitigating the pest's impact.

Author details

Aydemir Bariş
Plant Protection Central Research Institute, Ankara, Türkiye

*Address all correspondence to: aydemirbaris01@gmail.com

References

[1] Alkan B. Entomology of Agriculture. T.C. Ministry of Agriculture Ankara High Agricultural Institute Textbook. Ankara: A.Y.Z.E. Printing House; 1946. p. 156

[2] Bodenheimer FS. A study on insects harmful to agriculture and trees in Turkey and their control. Bayur Printing House: Ankara. 1958;346

[3] CABI, Distribution Maps of Plant Pests: *Myiopardalis pardalina* [Internet]. 2025. Available from: https://www.cabidigitallibrary.org/doi/10.1079/DMPP/20056600124 [Accessed: July 01, 2025]

[4] EPPO, *Myiopardalis pardalina* (CARYPA)–Distribution [Internet]. 2025. Available from: https://gd.eppo.int/taxon/CARYPA/distribution [Accessed: July 01, 2025]

[5] Barış A, and Çobanoğlu S. Investigation on the biology of melon fly [*Myiopardalis pardalina* (Bigot, 1891) (Diptera: Tephritidae)] in Ankara province. Turkish Journal of Entomology. 2013;**37**(3):293–304. https://dergipark.org.tr/en/pub/entoted/issue/5700/76200

[6] Barış A, Çobanoğlu S. Kavun sineği *Myiopardalis pardalina* (Bigot, 1891) (Diptera: Tephritidae)'nin morfolojik özellikleri. Gaziosmanpasa University, Journal of Faculty of Agriculture. 2014;**2**:72–79. https://dergipark.org.tr/tr/pub/gopzfd/article/95864

[7] Barış A, Çobanoğlu S, and Çavuşoğlu Ş. Determination of changes in tastes of İpsala and Kırkağaç melons against melon fly [*Myiopardalis pardalina* (Bigot, 1891) (Diptera: Tephritidae)]. Horticultural Studies. 2016;**33**(1):47–56. https://horticulturalstudies.org/abstract.php?id=53

[8] Toyzhigitova BY, Dinasilov SŁBKP, Zhunisbay A, Wołejko E R. Biological and chemical protection of melon crops against *Myiopardalis pardalina* Bigot. Journal of Plant Diseases and Protection. 2019;**126**:359–366. DOI: 10.1007/s41348-019-00231-x.

[9] Ashmead WH Description of *Pteromalus coloradensis* (Hymenoptera: Pteromalidae). Proceedings of the Entomological Society of Washington.1890;**1**:56–58.

[10] Vuillemin P. *Beauveria bassiana*, a fungal pathogen of insects (Cordycipitaceae). Bulletin de la Société Mycologique de France. 1912;**28**:145–150

[11] Sorokin N. *Metarhizium anisopliae* and its application in insect pest control. Memoirs of the Russian Entomological Society. 1883;**17**:55–64